Surrounded By Thunder:

The Story of Darrell Loan and the Rocketmen

By Tom Williams

Surrounded By Thunder: the Story of Darrell Loan and the Rocketmen

Copyright © 2013 by Tom Williams

Inspire On Purpose Publishing
909 Lake Carolyn Parkway, Suite 300
Irving, Texas 75039
(888) 403-2727
Website: http://www.inspireonpurpose.com/

Publishing Partner: Ascot Media Group, Inc.
The Woodlands, TX 77393
(281) 333-3507
Website: http://www.ascotmedia.com/

Printed in the United States of America
Library of Congress Control Number: 2013901849
ISBN 10: 0988753367
ISBN 13: 978-0-9887533-6-5

To book Tom for your next speaking event, or for media inquiries, please contact CaptTom@MarcoIslandToday.com.

Author's Note

Surrounded by Thunder is the story of a rocketman and a single account of one of the greatest endeavors of all time. It is the story of an extraordinary aeronautical engineer, his family, colleagues, and friends, and the historical narrative of a time in America that has never been surpassed.

All the chronological facts in this work are correct. The mission dates, the astronauts named, and many of the unforgettable characters that make up this account of America's race for space and then to the moon are accurate. This is however a historical narrative and written with the goal that it would read like an adventure.

Much of the dialogue in this work cannot be held as precise. Every account of a great achievement has a different point of view, and this narrative of the rocketmen in America has a supporting cast of thousands that can never be fully named or quoted with total accuracy. This is a story of a time not to be forgotten and a chapter in history that truly was and always will be, *Surrounded by Thunder*.

Reviews for: *Surrounded by Thunder* –The story of Darrell Loan and the Rocketmen

"Tom Williams has the tale-telling talent to mix tons of factual history with a gripping style more often seen in works of pure fiction."

"*Surrounded by Thunder* is beautifully crafted. It makes the reader eager to stay glued to the page, while yearning to look skyward, as if to see first-hand the events Tom brings to life so well."

--- Don Farmer, novelist, former CNN News anchor and ABC News correspondent and bureau chief.

A Different Perspective ...

Books about our heroes of flight, astronaut biographies and program histories are generally considered exciting and 'sexy' to readers smitten by the 'Golden Age' of aviation and space exploration.

But rarely do we get a true "behind the scenes" glimpse from the perspective of those unsung heroes who actually put our guys 'up there' - both into space as well as the stratosphere of publicity and gallantry. Every astronaut, from Mercury through the Space Shuttle, has always lauded those pioneers whose scientific genius enabled them to explore.

Tom Williams has done just that, giving us a glimpse and insight into one of those brilliant minds, a hugely causative factor in the space program's success. From America's first satellite, Explorer I, through Apollo and walking on another planetary body, engineer Darrell Loan had a hand in all of it.

Surrounded by Thunder ventures into this new realm, a highly informative and a true 'must read' for space program aficionados.

---Al Hallonquist, Aerospace Historian, Life Member - National Aviation Hall of Fame and the Flight Test Historical Foundation

"*Surrounded by Thunder* is a story of science-truth more exciting than any science-fiction tale. From his exhaustive interviews with one of America's pioneer rocket scientists, author Tom Williams delivers an enthralling account of how earthlings became spacemen in NASA's early days. This read is a real-life space thriller."

---Chris Curle, former CNN News anchor and print journalist.

Darrell Loan's thrilling experiences during the dawn of the Space Age read almost like a 'Forrest Gump' adventure. An Iowa farm boy grows into a key player behind scenes we know only from history books – the first American satellite launch, the pioneering

Mercury flights of Alan Shepard and John Glenn, the oft-overlooked but vital Project Gemini missions, culminating in the tragedy and ultimate triumph of Apollo. A true unsung hero, Darrell Loan witnessed history from a front-row seat, always *Surrounded by Thunder*.

---Roger Guillemette, aerospace journalist, former *Space Online* producer/reporter and *SPACE.com* correspondent

Surrounded by Thunder: the Story of Darrell Loan and the Rocketmen

By Tom Williams

As the Cape Kennedy countdown continued toward zero and the launch of the first manned mission to the moon, the words of a fallen president echoed though the minds of every NASA engineer, technician, and official:

"Those who came before us made certain that this country rode the first waves of the Industrial Revolution, the first waves of modern invention, and the first wave of nuclear power. This generation does not intend to founder in the backwash of the coming age of space. We mean to be part of it — we mean to lead it."
John F. Kennedy

"I believe that this nation should commit itself to achieving the goal, before this decade is out, of landing a man on the moon and bringing him home safely to the earth."
John F. Kennedy, May 22, 1961

"When Kennedy made that speech and set the goal for a man on the moon, no one can imagine how excited we were."
Darrell Alden Loan speaking about the top NASA technicians and engineers of the space program design team.

Surrounded by Thunder is based on a true story.

Chapter 1

"If I die, keep the program going. This is tricky stuff."

— Virgil "Gus" Grissom

When a full moon rises over a snowy Iowa landscape, the passage of time slows down. The lunar body seems bigger and brighter as it breaks free of the snow, and as the celestial mechanics that drive the universe lift the moon higher, the folks from Iowa can see there are mountains on the moon, and even a sandy ocean called the Sea of Tranquility.

Darrell Loan was born in 1929 in Iowa City. Many have said when the going gets tough the tough get going, but when Darrell came into the world the tough was going to get about as hard-hitting and rough as anyone could imagine. The Great Depression was on the way and as every child learned growing up during the 1930s, life was not what it used to be.

The hints were all around for a youngster to discover, and it was easy to listen as the older folks talked about the days before the Stock Market Crash and the good times after the Great War. As Darrell listened to the stories of the 1920s, there was little doubt that the boisterous tales of prosperity and yesteryear were different from what the world was like now.

The fact that times were tough was always on the doorstep with the Sunday paper, or in whispers down at the drugstore, but when Darrell heard about the bank failures and his own family it was a hard lesson for an eight-year-old to understand.

Darrell's dad, Charlie Alden Loan, unlike many, did have a job, and there was always food on the table, but the rumors of what might happen if things got worse were a hardship for an eight-year-old, especially since Darrell had what the folks in Iowa called imagination.

Charlie Alden Loan was Irish and Martha Marie Heldt was German. They were both Americans, but Charlie could remember his grandparents speaking with an Irish accent, and Martha Marie could easily speak German. They fell in love as Americans do, and began to build their lives in Iowa.

The land was good, family values were strong, and it was hard to understand when the bank failed and took all the money. The trouble started in a far-and-away place called New York City and left a scar that a generation would bear and could never forget.

Charlie Alden had more than enough cash in the bank to cover the mortgage, but when the banks crashed, everyone lost all their savings. The hard part to understand was when the same bank that lost all the money wanted the mortgage paid. The bankers even sent the sheriff out to get the diamond wedding ring on Martha Marie's finger, but Charlie Alden explained to *that* law officer — man-to-man — that just wasn't going to happen.

Over the years, Martha Marie and Charlie Alden rebuilt their financial lives and grew stronger. From that fateful day on, however, Charlie Alden never trusted a bank, and cautioned anyone who would listen to always pay cash and never to finance.

The Iowa home was always full of music as Martha Marie could play anything on the piano, and their lives were full of lessons that Darrell would always remember.

Every few weeks it seemed there was an engineering lesson that needed to be learned. The Charlie Alden Loan family had a pony, a barn, and a latch on the barn door that wasn't pony-proof. The fact that the pony was smart was not a question, because every so often the pony would get bored and venture into town. She would always sneak out when no one was watching — sometimes in the middle of the night — but always when it was least expected. The results of course were always the same. The phone would ring; Charlie Alden would answer, and learn that his pony was standing in the middle of the road, watching traffic at a red light in downtown Iowa City. After the news, Charlie Alden and Darrell would hop into the family Pontiac and drive into town. They would collect their pony with a halter and a rope, and as the Pontiac slowly cruised back to the farm, the pony would trot happily alongside. The mystery of the clever pony named "Typical Beauty" and the failing barn door latch would be a lesson for the young Darrell Loan and the beginning of an engineering career second to none.

When the phone rang and the news came that the pony was out on the town, Charlie Alden would scratch his head. "One of you kids must have left that latch undone. There's no way that pony could get out by herself."

Darrell knew that he was not responsible for the misadventures of the pony, nor were any of the other kids, but he had a strong suspicion there was something wrong with the system that held the pony in place.

Typical Beauty had learned to lift the latch with her nose anytime she wanted a little adventure. After several trips into town, with the halter, the rope, and the Pontiac, Darrell learned that the latch needed to be on the other side of the door. The solution was simple, but it was the beginning of a trademark and style of engineering that would land a man on the moon.

Chapter 2

Darrell liked basketball. He liked everything about it. He liked the teamwork, the excitement, and even the smell of the gym after it had been empty and the lights came on. In his Cosgrove, Iowa, high school, Darrell Loan was a basketball star. He perfected a technique of passing the ball back and forth between teammates so that dribbling would not slow everyone down. It worked perfectly, the Cosgrove team won, and Darrell was awarded a full college scholarship for basketball when he graduated from high school in 1947.

As exciting as college was, and as enjoyable as it was to play basketball for the University of Iowa, there was a dark specter gathering on the horizon that was going to be trouble. Darrell could remember all too well, the day in December that would live in infamy, when he was listening to the radio with Charlie Alden and Martha Marie. On that fateful Sunday when Pearl Harbor was bombed, and America came into the Second World War, there was an escalation of science and technology that would bring the world into a new era.

Five years after the defeat of Japan and after two years of studying accounting and many victories in basketball, the Korean War began and once again global events dominated the radio airwaves and the Sunday newspapers.

The Korean War began on June 25, 1950, when the Chinese- and Soviet-backed North Koreans flooded across the 38th parallel to swallow up South Korea. The plan was to act before any other nation could rally and send support against the Communists. The result was the beginning of the Cold War with America and the United Nations allies fighting for the freedom of the South Korean anti-Communists. The conflict rapidly escalated into an all-but-nuclear confrontation as the newest Soviet and Chinese military hardware was tested against armaments largely produced by the USA. It was the beginning of jet fighter warfare and another step toward futuristic aviation and the conquest of space.

When the Korean War began, Darrell closed the accounting ledgers, took a last shot from the basketball foul line, and joined the U.S. Army. After boot camp and a year of officer's training school in Fort Hood, Texas, the 2nd Armored Division known as "Hell on Wheels" arrived in Korea. It was muddy, bleak, and hideous, and nothing could prepare any American for the horror of the relentless conflict.

For Darrell, music was mathematics and math was musical. Martha Marie could hear a song once and play it on the piano and her son was blessed with the same ability. He was also able to read music with only one lesson, and the tenor and soprano saxophone came to him as easily as the piano came to his mother. Sight reading music was as easy as speaking and all during high school and college Darrell played the saxophone in every band he could find. The memory of the music that was Glenn Miller and Benny Goodman was the personal *Moonlight Sonata* that held a young man from Iowa together during the horrors of Korea.

"We were overrun twice," Darrell explained as he recalled his service in Korea. "I watched my best friend as he was cut in half with an 88-millimeter shell."

"Overrun" in battle is a term company commanders don't want to think about. It happened to George Custer in the battle of Little Big Horn and it happened to the Americans in Hawaii when the Japanese attacked Pearl Harbor. Overrun simply means overwhelmed with enemy forces.

When situations deteriorate in armed conflict, battlefield promotions are awarded to those who stand out as leaders. After several months in the far away Orient, fighting everything the Chinese and Soviet Communists could muster, Darrell Loan was promoted in the battlefield to First Lieutenant and then to acting Captain and ultimately to Company Commander.

During the heat of battle when the threat of disaster becomes more of a reality than a possibility, heroes rise to the occasion and save the day. The 30-caliber machine gun was a very effective weapon during the Korean conflict, but when fired continuously, the barrel of the fully automatic weapon becomes heated to the point of melting. Replacement barrels are attached when the barrel in use becomes red hot, and asbestos gloves were issued for whenever the near-melting barrel needed to be unscrewed and changed.

"Once when we were desperate and under heavy attack," twenty-two-year-old Company Commander Loan explained, "our 30-caliber machine gun was really working overtime. We were warned about the barrel becoming overheated and exploding. When there was no time for the asbestos gloves, one of our soldiers grabbed the red-hot barrel with his bare hands. He twisted it free and saved our lives. The barrel was glowing bright red, and it burned the flesh of his hands right down to the bone. There was no doubt the man was a hero — his hands were ruined — but that was just one example of what happened almost every day."

On July 27, 1953, the United States, North Korea, and China signed an armistice and the fighting was over, but technically, the two Koreas are still at war. The peace process remains fragile, and, to this day, the North and South Korean borders remain the most heavily patrolled and fortified in the world.

After two years in the battlefield fighting the Communists, Darrell returned stateside and was approached by his new commanding officer.

"Loan, I know your time is up and you're waiting for a discharge, but I want you to sign up for the reserves. But I *do* have to tell you — we can call you back at any time — fighting these Communists is far from over. Everybody knows that the Russians will be next."

Without hesitation Darrell responded, "No sir, no thanks. I've done my time, and I want out."

"Loan," the officer glowered, "I swear to God, if you don't sign up and join the reserves, I'll bust you all the way back to an enlisted man!"

For Darrell, the horrors of Korea and two years on an oriental battlefield were enough. He refused to be bullied back into the reserves, but, true to his word, the imposing officer took away the acting Captain rank that served as Company Commander and the battlefield promotion to First Lieutenant, and reduced the short-time soldier from Iowa to corporal.

Chapter 3

"If you want something done, ask a busy person to do it. The more things you do, the more you can do."

— Lucille Ball

After army boot camp and during officer's training school, Darrell met his future wife Audrey at a community dance in Iowa City. The romance was a fast-paced whirlwind with Korea looming quickly. Darrell and Audrey were married on the same day, with another couple, in order to share and save on expenses, in 1951.

When Darrell returned to Audrey and the University of Iowa, he discovered to his vast disappointment that he was now too slow to qualify for a basketball scholarship. He also discovered that an accounting degree was not challenging enough and switched his major to mechanical engineering with an aeronautical option. Watching the fighter jets in Korea had made an impression.

Paying for school without the scholarship was another problem, so he picked up the saxophone and began playing in a traveling dance band all around the Midwest. Every dance band performance produced money for college and for the next four years, Darrell reveled in the music from the Big Band Era. There was very little sleep along the way as playing saxophone until two o'clock in the morning and driving back to Iowa City captured most of the night, but the lack of sleep and the pressure to perform both in class and on the bandstand were just another lesson in life and another preparation for what was to come.

The University of Iowa has always had a huge rivalry between the students that chose engineering and the students that chose to study law. Every graduating year there was some type of prank or incident that the future lawyers would play on their engineering counterparts, but for the graduating class of 1957, the up-and- coming legal counselors for the Midwest really did their homework.

Springtime in Iowa often comes late, but when Darrell and Audrey Loan were watching the daffodils rise and the spring planting of corn and soybeans spread across the countryside, the almost-lawyers were planning a stunt that would stagger the imagination.

On the night of the Engineering Department's University Mecca Ball, the Union Hall was filled with young ladies dressed in their finest gowns and the senior engineering students proud as peacocks as they waltzed their girlfriends and brides across a highly polished dance floor. The evening was perfect, as spring was in the air, and a full moon on the rise. The band was the best ever, and there was even one carefully arranged song from a Tennessee boy named Elvis Presley. There was a big punch bowl and crystal cups, there were cookies and fancy snacks, and there was enough confidence in that graduating class of 1957 to reengineer the world. The future engineers from the Midwest were having a blast, but the almost-lawyers from across the campus were about to make all of that change.

The first half of the Mecca Ball was flawless. The young women could not have looked prettier in their delicate finery and the tuxedos of the men—although begged, barrowed, and rented—were as neat as a new pin and as clean and spotless as a fresh sheet of drafting paper.

When the second half of the Ball began and the dance music started, the lawyers-to-be began the sneak attack.

With the Union Hall once again filled with dancing ladies in beautiful gowns and the young engineers concentrating on the project at hand, it was more than a shock when the first greased pigs took to the dance floor. They were young pigs from a nearby farm, full of energy and springtime, and they had all been thoroughly smeared with axle grease. At first, it seemed, there were only a few, and the dutiful engineers abruptly dropped the hands of their dance partners and tried to be heroic, but the more they tried to catch the running and squealing pigs the more slippery and frantic the pork bellies became. As the gala Ball that began so beautifully began to explode with the screams of the young ladies, the squealing pigs, and the curses of the young engineers, the almost-lawyers hidden in an adjoining room began to giggle and release more greased pigs. By the time the dance band stopped playing and the majority of the engineering girls were lifting their skirts and heading for the doors, there were twenty greased pigs running around the dance floor. Soon, it was apparent that the law students had struck, the second half of the ball was over, and the young ladies were definitely not happy to have axle grease all over their specially chosen gowns. It was also a time when a group of engineers stood outside the Union Hall under a full moon, smoked cigarettes with grease on their hands, and decided that the future lawyers of the Midwest were going to get a carefully engineered retaliation.

During the following days of the greased pig fiasco, three engineering seniors drafted a plan worthy of genius. The almost-lawyers were going to pay dearly for the ruined Ball, and Roger Wymore, Jim Hack, and Darrell Loan were going to make certain the engineering vengeance would go all the way to the top.

The Dean and the most recognized head of all the lawyerly was going to be the target because in the almost-lawyer world, everything came down from the top and rained on the heads of the underlings.

The fact that the weather was good and the campus was busy was a good thing, the three engineers decided. It was almost graduation time, and everyone was outdoors and walking the college grounds. There were strangers around looking for the most talented students with the best grades, but as everyone knew, strangers could be blamed for almost anything. Most of the strangers were searching for the very best from the University of Iowa to ensure the future of American industry, but no one could know in 1957 that American engineering would take a journey to the moon during the very next decade.

The plan for the engineer's revenge began with a good idea and then got better. Outside of town, there was a rodeo man. He was known throughout the Iowa City area as an entertaining fellow, and from time to time, he would host local rodeo shows. The rodeo man had a ranch with all the animals the old Wild West had to offer. There were bucking broncos with horseshoes of steel, there were fiery-eyed bulls with evil looking horns, and there was even a jackass that was particularly stubborn and perfectly designed for revenge on the almost-lawyers.

With only a week before graduation, Roger, Jim, and Darrell drove out to the rodeo ranch. Roger was driving, Jim sat in the middle, and Darrell rode shotgun as he contemplated the plan and the springtime countryside.

"We need to borrow your jackass," Darrell explained to the rodeo man. "We promise to take good care of him, and we also need to borrow your truck to haul him out to the campus."

The exact arrangements for the leasing of the jackass remain to this day a mystery, but it was clear that the rodeo man didn't think much of lawyers. He did, however, have a good sense of humor and was more than ready to help when he heard about the greased pig ball and the sneaky approach to sabotage the engineers and their lady friends.

On the day of Operation Jackass, Roger Wymore went across campus and tried to look lawyerly. He was carrying books as he walked through the doors of the law library that housed the offices of the Dean of Law, and when he crossed over the threshold, the campus clock chimed four o'clock. The law library and the entire lawyerly building closed up tight at five o'clock sharp.

After a few minutes deep in enemy territory, Roger found himself walking along the bookshelves in search of a hiding place. All of the almost-lawyers were busy and no one saw Roger turn a corner and slip into a cleaning closet. When the muffled chimes of the campus clock struck five o'clock, there was a single engineering senior alone in the hallowed halls of the future litigators.

Meanwhile, Darrell and Jack were waiting for nightfall. The rodeo man's truck was loaded with the jackass, but dropping a tailgate and pulling a stubborn old mule out of the back of a worn-out Dodge was not something that was ordinarily done in the daylight hours on campus. Normally, the only type of jackass that ventured into the law library walked on two feet.

When the sun finally disappeared over the western plains and darkness covered the University of Iowa, Darrell popped the clutch on the battered old pickup and eased Operation Jackass into place at the entrance to a side door of the law library.

After turning the jackass around—because a jackass will not walk backward—Darrell and Jack led the old donkey down a ramp and into the unlocked and opened door where Roger was waiting.

After only a few steps into the sacred ground of the law library, the jackass had an attack of diarrhea. The old donkey was nervous as he was led up the polished marble steps and the diarrhea continued all the way up to the second floor and into the office of the Dean of Law. The steps, the floor, and the Dean's office were a smelly and slippery mess by the time the jackass was tied to the Dean's desk, and as Roger, Jack, and Darrell bolted down the stairs, they had no idea that two security officers were waiting by the door. Just before the engineering seniors were ready to make their escape into the hands of the campus police, Roger saw the searching flashlight beams and signaled for an abrupt halt. After a few tedious moments with thoughts of expulsion, betrayal, and failure racing through the plan, the three engineers watched the security guards continue their rounds. To this day, no one understood why the jackass waited to have diarrhea until he was inside the hallowed halls of the almost-lawyers, but many believe it was an exchange of professional courtesy.

When the library opened the following morning and the janitorial staff saw what was waiting, telephones began ringing all over campus. It wasn't long before the jackass was connected with Iowa City's rodeo man and the SPCA found out there was a sick old donkey tied up in the executive offices on the second floor of the law library.

When the Dean of Law and the campus officials confronted the rodeo man, he was ready for the chief litigator and his almost-lawyers.

"There was two old desperados came out to my ranch with big guns. They were rough customers." The rodeo man lied. "They told me they wanted the jackass and the truck, and if they didn't get what they wanted that was all for me."

"Why didn't you call the Sherriff and tell him what happened?" the Dean of Law demanded. He was an overweight and surly-looking man that always wore a severe and tight-fitting black suit.

"Because they told me all they wanted was to borrow the jackass and the truck, and if I made a fuss with the law, there would surely be trouble. They warned me they were coming back, and I had better be quiet."

Although the rodeo man stuck to his story and kept his word to the engineering seniors, the Dean of Law wasn't convinced, and launched a campus-wide investigation. Lawyers love investigations almost as much as they love out-of-court settlements.

"There are clearly engineering students," the scornful Dean announced to reporters from the campus newspaper, "that are behind this obvious prank. Once these individuals are identified and arrested, they will *not* graduate from the University of Iowa."

The investigative storm that followed Operation Jackass raged though the following week and was made even more intense because of the cost involved in removing the jackass from the office of the Dean of Law. A jackass will not walk backward and will not be led down two flights of stairs. The stubbornness of mule-like animals is why the young engineers chose the jackass. But not in their most hopeful imaginations could Roger, Jack, and Darrell have known that the Society for the Prevention of Cruelty to Animals insisted that a second story brick wall be taken down and the jackass lowered to the ground with a full body sling lifted by a crane.

With seemingly the entire student body and many of the professors watching, the Iowa City rodeo jackass was lifted out of a big hole on the second floor of the University Law Library and carefully lowered to the manicured grounds. Every engineering student was present and all of the almost-engineers were smiling.

Chapter 4

"I have never let schooling interfere with my education."

— Mark Twain

With the jackass hauled back to the rodeo ranch, and the investigation into the perpetrators of the lawyerly hall toned down to a dull roar, Roger, Jack, and Darrell began to believe that they might actually graduate. There was a lot of worrying about the prank, and more than a few sleepless nights, but to complicate circumstances further, there were strangers on the campus looking for Darrell Loan.

At first, Darrell and the other members of Operation Jackass thought the newcomers were actually Pinkerton men helping with the lawyerly investigation, but it soon became came apparent that one of the strangers dressed in a new suit was from Bell Laboratories. The recruiter from Arizona was apparently impressed with Darrell's grades and his record in the military. The goal and the job for the man from Bell Labs was to engage and sign up as many of the top engineering seniors as possible before heading back to the Southwest. Bell Labs was determined, and they wanted the best.

The idea of summer in Arizona, however, did not sound appealing, and when push came to shove and the man from Bell Laboratories was pitching hard, Darrell explained with his characteristic nonchalance, "I'll just have to think about the offer. Thanks for asking."

Within days of graduation, another stranger in a sharper suit approached and announced that he was from Great

Neck Long Island in New York. He also explained with a New York accent that he was from the Sperry Corporation that made gyroscopes. He also confidently made clear that Sperry had developed the bombsites that were so amazingly accurate during World War II.

The aerial viewing devices perfected by Sperry and installed into the B-17 Flying Fortress were directly responsible for the incredibly precise bombing raids that destroyed Nazi Germany. Sperry was a mover and shaker in the up-and-coming world of advancing aircraft, and the recruiter from New York was confident the aeronautical option in Darrell's engineering degree was the perfect fit for the Long Island corporation.

"I'm not sure I'm cut out for what you want," Darrell confessed to the man from New York.

"Listen, kid, with your grades and five years of engineering school, we can teach you anything else that *we* want you to know." The New York accent sounded strange in the Midwest, but it also sounded successful, confident, and like something right out of the movies.

"How about signing up right now and packing your bags? We'll get you and the little missus straight out to Long Island."

The money that New Yorker offered was more than Darrell had ever hoped for because he was just about as poor as a church mouse. Charlie Alden and Martha Marie had always provided a good home with food on the table, but there wasn't any cash left over for anything but the basics.

What the New Yorker explained quickly *after* the papers were signed was that to get out to Long Island, Darrell would have to pay to get there. This was because, like many big corporations, the executives in charge knew that if a recruit had to pay their own way, they would be less likely to get homesick, give up, and go back home. The deal was firm, the Sperry man explained, "Pay your way out, show up on your own, or no job."

Despite all the tension over Operation Jackass, the new job at Sperry, and the worry over how to come up with the money to move Audrey to New York, Darrell and his classmates graduated from the University of Iowa in 1957.

The move to New York was not easy. At first, Darrell went to the bank in Iowa City and asked for a loan. The bank manager explained that a loan to get to New York City was just not possible. Banking loans were for farming, building, and livestock. That was what was important. There was no money available for something as unheard of as moving to New York.

After hearing the news, Charlie Alden drove with Darrell back down to the bank and had a private meeting with the bank manager. The meeting was not all that private as everyone in the savings-and-loan department could hear the muffled shouts of Charlie Alden as he ranted and raved and officially secured the money necessary for Darrell and Audrey to move to Long Island. Two weeks after graduation, Darrell and Audrey flew in to New York City.

After arriving in Long Island, Darrell met with the officials at Sperry.

"We want you to take a crash course in electronics," a terse Long Islander explained. "You'll be finished before October."

With Audrey moved into a rented home in nearby Levittown, Darrell began to learn about electrical circuitry, capacitors, transformers, and something new to replace vacuum tubes called transistors. The classes were indeed a crash course with one-hundred-and-ten new Sperry recruits studying very hard until the end of summer.

On the Friday before Labor Day, all of the students knew that something was wrong. When class began, the lead instructor merely called out the names of six individuals and made a simple announcement: "Everyone that was called is required to pick up your books and get out of here. You also need to report back to this office on Tuesday morning."

Over the weekend, Darrell learned that the remaining Sperry recruits—over one hundred—that had not been called—had been dismissed. Darrell Loan was the third name called out of six. The following Tuesday, the surviving six were awarded with electrical certification documents and assigned troubleshooting duties all across the eastern seaboard.

After no further training or discussion, Darrell was sent out on the road in a 1957 two-door Mercury station wagon. The Mercury was red and brand new. His newfound assignment was to solve every problem that the military was having with any product made by the Sperry Corporation. After only a few weeks, it was obvious the new recruits were covering areas that were previously serviced by six individuals.

The new Mercury was soon traveling from Norfolk Naval Base in Virginia for the Navy, to Cherry Point Naval Air Station for the US Marine Corps, or to any military installation in between where aircraft or guidance systems were under scrutiny. There were problems with Sperry-made engine analyzers, auto pilot systems, and compass systems. Wherever there was a problem with any Sperry product, the Sperry troubleshooters were assigned to the field.

There were other assignments waiting for the specialty engineers that were right out of the movies— particularly the 1950s science fiction films—and some of those projects were coming from something that everyone was beginning to call "The Atomic Age."

Chapter 5

"Oh Little Sputnik flying high, with made-in-Moscow beep, you tell the world it's a commie sky and Uncle Sam's asleep . . ."

— G. Mennen Williams

On Friday October 4, 1957, the Soviet Russians were throwing a party. They had a good reason. The reception at the Soviet embassy in Washington was apparently to encourage discussion for the International Geophysical Year and the previous week of meetings regarding the most recent developments in international scientific achievement. Dr. John P. Hagen was in attendance as a senior scientist with the Naval Research Laboratory as he was in charge of America's fledgling attempt to launch an artificial satellite.

With the embassy party well under way and Dr. Hagen in a deep conversation with a group of Soviet scientists discussing the possibility of a Russian-launched satellite, the room was suddenly galvanized when Walter Sullivan, a reporter for *The New York Times*, received an urgent message from his editor. After making a phone call and quickly speaking to several members from the press, it was confirmed that the official Soviet news agency had announced to the world that a 183-pound, Russian-made satellite named "Sputnik" was now orbiting the earth at a height of nine hundred kilometers. Every ninety-six minutes the Soviet Sputnik was crossing over the United States.*

Dr. Hagen felt his world tilt and hopes for a U.S. naval victory shatter as he considered the first American attempt at spaceflight was to launch a satellite that weighed less than four pounds, while Sputnik was already aloft and circling the globe with a physical weight of almost two hundred pounds.

To make Dr. Hagen's and the American efforts seem even more hopeless, the following month, "Sputnik-2" was launched from the desert-based rocket testing facility in the Kazakhstan region of Soviet Russia. This time the Communist spacecraft weighed 1,120 pounds and carried the first living creature into outer space. A dog named Laika rode in the Russian rocket and lived until the oxygen gave out, but Sputnik-2 carried the haunting remains of the Russian dog circling the earth for two hundred days.

With the launch of Sputnik-2, the Eisenhower administration announced at a White House press conference that America would soon enter what was rapidly being called the "Space Race." Dr. Hagen's U.S. Navy Vanguard rocket booster was to launch the United States' first satellite into space on December 6, 1957.

After the news of the Communist spaceflights, Americans scanned the night skies for a threat they could often see as Dr. Hagen and his team rushed to make ready the Project Vanguard booster and the United States' first attempt to touch the edge of space.

The Russian satellite technicians polished the round metallic body of Sputnik to shine and reflect light so it could be seen by the American people as it crossed over the United States, but what most Americans saw streaking overhead — because Sputnik was so small — was the second stage of the Sputnik rocket booster as bright as a first magnitude star.

Chapter 6

"A pessimist sees the difficulty in every opportunity; the optimist sees the opportunity in every difficulty."

— Winston Churchill

Because of the Sputnik launches and the continuous threat of long-range and meticulously planned Soviet aggression, President Eisenhower decided that America needed to be ready for anything. During 1949, the Russians exploded their first atomic bomb, and intelligence sources warned that much stronger hydrogen weapons were under construction somewhere behind the Iron Curtain.

The Soviet Russians rolled tanks into Budapest in 1956 after the Hungarians dared to defy the crushing Communist rule, and during the same year, freedom of passage for the Suez Canal was threatened, and the entire world was on the brink of atomic war. Terms like "nuclear fallout" and "gamma radiation" began to settle into everyday conversation even in places like Iowa City.

The Federal Civil Defense Administration was recommending that everyone in the United States prepare for atomic attack. Children in school were taught to "duck and cover" under their desks at the sound of a wailing air-raid warning, or at an unexpected sneak attack when the bright-white-flash of an atomic explosion came through the windows of the classroom.

Individual families were encouraged to build fallout shelters with at least three feet of earth covering the basement-like emergency dwellings. With civil defense teams armed with Geiger counters to measure radioactive fallout particles, the fear of atomic warfare and the end of everything normal was a new way of life for America during the 1950s.

As the two-door Mercury cruised along the eastern seaboard, and threats real or exaggerated came across the radio broadcasts, it was obvious the world was in trouble and changing fast. Escalating even faster were the problems that the military was having with Sperry. Quite often, the troubleshooting dilemmas were highly complicated, because if they were not, the Naval and Marine engineers would have already solved the issue. On many occasions, the problem at hand was urgent and made even more pressing by the increasing conflict between the USA, Red China, and Soviet Russia. Because of delicate political climates, and because of the rapidly growing fear of communist espionage, strangers once again began to appear at the University of Iowa and in the countryside surrounding Iowa City.

The nondescript strangers always arrived in pairs, and they began to ask questions. At first, they approached Charlie Alden and Martha Marie. They had a list of questions and they began to probe deeply.

"Has your son ever mentioned having any friends from Eastern Europe? Has Darrell ever talked about having sympathies for East Germany, Russia, or Red China? Would there be any circumstances that would make a boy from Iowa unpatriotic? Does your son harbor any ill feelings for the U.S. Army because of his time served in Korea? Have you ever seen any letters or correspondence from any foreign country?"

All of Darrell's friends, teachers, and relatives were approached, and always the results were the same. The strangers in suits were from the FBI, and as they probed even deeper into the past, and even into Audrey's family and friends, the answers were always identical.

"No, Darrell and Audrey do not know anyone from Europe, Eastern Europe, Russia, or China, and they are certainly not unpatriotic or Communists. No, we have never seen or heard about any letters with foreign stamps!"

There are four classifications of secrecy in the United States: Confidential, Secret, Top Secret, and Crypto. The designation of "Crypto" is the absolute highest level of secrecy in the American government.

Sperry chose Darrell because of his grades. He was evaluated during the electronics crash course, and he made the top six out of 110 carefully selected students from all across the USA. As far as the Sperry Corporation was concerned, and more importantly the FBI and the U.S. military, there were extremely sensitive assignments that could be entrusted to only a few. At the age of twenty-six, Darrell Loan's secrecy classification was elevated to Crypto, and at a secure building at the Norfolk Air Station, he was given the task of arming nuclear weapons.

In 1957 there were a series of relays that had to be aligned in a perfect sequence for a nuclear weapon to detonate. Without the perfect sequence, the bomb would not reach critical mass and would not create an atomic explosion. The result of a faulty relay sequence would be a small explosion of TNT effectively releasing poisonous plutonium.

Because of the secrecy involved, the sequence of the relays had to be memorized. There were no notes, no instructional manuals, and absolutely no one who knew the sequence code that did not have the secrecy classification of Crypto. The only possible way an American atomic bomb could be armed in 1957 was by someone who had memorized the correct relay progression.

After several days at the Norfolk Naval Air Station surrounded by the military, Darrell was on the road again — his head reeling with sequence numbers — but this time he was headed for the Norfolk Naval Base and a big change in plans.

Chapter 7

"There is no doubt in our minds that Nasser, whether he likes it or not, is now effectively in Russian hands, just as Mussolini was in Hitler's."

— British Prime Minister Sir Anthony Eden speaking about the Suez Canal

"We have some serious problems with some of your equipment," Captain Houston explained. The Staff Captain was walking with the newly arrived Sperry consultant along the aircraft carrier docks. Moored along the waterfront were the towering walls of gray steel that were a task force of ships preparing for sea. Houston paused to examine the bustle of dockside activity and then led the way through a column of rumbling forklifts and into the relative quiet of an aeronautical workshop.

After the hanger door closed and Darrell was alone with the naval officer, Houston produced a Sperry manufactured guidance system for the A-4D attack aircraft. The object under workbench lights was a rubber-sealed black box that looked unfamiliar. It was obvious however that the compass system was sealed and that the naval engineers were concerned about breaking the seal and damaging the interior components.

Darrell had never seen the guidance system before, but, after only a minute, he shook the unit and heard loose parts rattling on the inside. With Captain Houston watching, the seal was broken, the hiss of a vacuum escaped, and the Sperry unit was opened on the workbench. An unattached circuit board fell out with a spaghetti tangle of multicolored wires as several small screws rolled out under the lights.

"Here's the problem," Darrell explained. "The screws came loose and the circuit board moved and shorted out. I don't understand how this could have happened."

"It only happens with the units that have been in use, and only with the units that have been on special maneuvers. The new ones work just fine. Only the planes that have been in training and have practiced the Idiot Loop have had the failures."

"The Idiot Loop?" Darrell asked.

Captain Houston lit a cigarette, sat on a stool at the workbench, and picked up the broken unit. "You have a Crypto security clearance? That right?"

"Yes, but you know that," Darrell looked the officer, "or we wouldn't be here having this conversation."

Houston blew a smoke ring. "The Idiot Loop happens in training, when the A-4D flies in to deliver an A-bomb. The plane comes in hot, launches the bomb, and then rolls over and flies out on the same heading—only upside-down and even faster. That's the only way the pilot can get away fast enough . . . before the explosion."

"That's it," Darrell cracked a grin as he suddenly understood. "They're pulling too many 'Gs,' the screws are coming out, and the circuit boards are failing because they get loose."

"Can you fix them?" Houston asked, "The units we have here—without getting new parts?"

"Sure," Darrell's smile was contagious. "A little blue-glip glue on the screws and they should be as good as new. They'll hold during the Idiot Loop, but the units won't be permanently sealed against the salt air. That's what the rubber seal is for. Rubber seals and O-rings are always tricky and delicate, but very important."

"Can you do this on all the planes that have the failed units?"

"As long as the circuit boards are not damaged, but they really should be sealed at the factory for long-term use."

"There's no time for that." Houston shook his head. "Have you heard about the latest at the Suez Canal?"

"About Nasser, Eisenhower, and the Russians?" Darrell suddenly didn't care for the direction the conversation was headed, and he definitely didn't like the look on the naval officer's face.

"Yes sir, that's the news, and now you're part of it." Houston leveled his gaze.

"What are you talking about?"

"Loan, you're *now* officially under orders —" Houston explained as he offered a cigarette, "Top Secret orders. You're going aboard an aircraft carrier headed for the Middle East, and you're going to fix all the failed units aboard all the A-4D aircraft. The carrier sails at midnight and you're going to be onboard. We've already contacted the folks at Sperry and they've agreed that if you can fix the problem, you're the man for the job."

"But my wife," Darrell said, "I'll have to call my wife. She won't like this — not one little bit. How long will I be gone — how long at sea?"

Houston's friendly manner had sailed with the tide. "This might be your first overseas assignment with a Top Secret security clearance, but there will be no phone calls. No calls to the missus, no calls to Sperry, and no letters or notes to anyone. This is a Top Secret mission to protect free passage to the Suez Canal and now you're part of it. You leave at midnight."

Before Darrell could speak, the naval officer raised his voice and called to the hanger door: "Ensign!"

Obviously, the Navy had already made plans. After the summons, a young man about the same age as the new troubleshooter from Sperry appeared at the hanger entrance and then saluted. He was wearing the crisp whites and cap of the U.S. Navy, and his rigid response to Houston's command was a perfected: "Yes sir," as he stood at attention.

"Ensign, as we talked about earlier, you are to escort this civilian aboard the carrier and show him to the assigned cabin. No conversation will be necessary."

Captain Houston offered his handshake, "Loan, thank you for once again serving your country, but I pray to God we won't need what you've been sent here to fix."

Within the hour, Darrell Loan, Sperry Corporation employee with a beyond Top Secret, Crypto security clearance, was aboard an atomic-armed aircraft carrier and shown to the A-4D aircraft hanger deck. With the bustling activity of a naval vessel preparing for sea all around, Darrell began dismantling the failed compass and guidance systems in all the fighter-bombers that had performed the "Idiot Loop."

After three days at sea at a classified location somewhere in the mid-Atlantic Ocean, a telephone rang at the home of Darrell and Audrey Loan in Levittown New York.

Chapter 8

"Since knowledge is but sorrow's spy, it is not safe to know."

— *William Davenant*

Audrey was frantic. It had been three days since she had heard from Darrell, and the tension and worry of what might have happened to her missing husband was driving her crazy. Repeated calls to Sperry revealed nothing but frustration as the inquiries into her husband's disappearance were only answered with mysterious suggestions about a storm at sea. Calls to Iowa were equally upsetting as no one at home knew anything about the abrupt lack of contact.

Ordinarily, Darrell would phone home when he was expecting to be late, or if the Sperry assignment at hand was going to take longer than a day, but not since the nightmare of Korea had Audrey been beside herself with worry. She knew that the work at Sperry was related and linked to the military, and that some of the work was secret, but after all the questions from the FBI, and the embarrassing investigations into her and Darrell's past, there was no way of understanding what was really going on. Three days without any word was beyond what she could stand, and Audrey was beginning to consider that maybe even Russian spies had kidnapped her husband. After all, every movie that everyone went to see had something about secrets, Communists, or worse, and the concept that *no one* knew what was happening was beginning to take shape.

On the morning of the third day, just when Audrey was ready to call the FBI, the telephone rang.

"Is this Mrs. Darrell Loan?" the voice on the phone began.

"Yes, yes. This is Audrey Loan."

"Mrs. Loan," the voice continued, "this is a special message only for you. I am working with the U.S. government, and I have been given the duty of calling and requesting your help. I have been authorized to explain that your husband Darrell is away on a special assignment and may not be able to contact you for several days. It is vital that you do not alert any local police regarding his absence, or anyone else. Your husband's work or his absence is not to be discussed with anyone. Do you understand everything I have told you?"

"Who is this?" Audrey was suddenly furious. "What is your name?"

"My name is not important," the voice continued. "What *is* important is that you tell me that you understand everything that I have said. No mention of your husband's work or his absence is permitted. Please tell me that you understand."

"I understand that this is crazy," Audrey sobbed. "I have been worried sick for three days and now someone calls that won't even give his name?"

"It is unfortunate that a certain amount of time had to pass before we could make this call, but *you* must know that your husband's work is *sensitive*. Do you now understand everything I have told you?"

"I understand that my husband has been missing for three days, he will be away for more than three days, and I am supposed to keep my mouth shut because his work is *sensitive*?"

"That is correct. Goodbye."

"Wait a minute!" Audrey's anger resurfaced. "What about me? Are my feelings not *sensitive*?"

When she realized she was talking to a dial tone, Audrey slammed down the big black New York telephone and began to cry.

Afterward, and later in the day, Audrey Loan went shopping. She wore a bright yellow dress and her best shoes and nylons, but as she walked along the shops in Levittown, she kept looking over her shoulder. There were men in suits of course, because most men wore suits, but some of the suits looked a little too new. What was really strange was that a couple of the men seemed to be watching her—and not in a nice way. With only one parcel tucked under her arm, Audrey went home, but all along the way she felt like she was in some crazy Alfred Hitchcock movie. A movie where her husband was a spy or something, but what that *something* was, she had no idea.

Meanwhile across the Atlantic Ocean, onboard an aircraft carrier loaded with jet fighter bombers and atomic weapons, Darrell stood on deck to watch the waves go by and wondered what Audrey was going to say when he got back to Levittown. All the A-4D guidance systems had been repaired. The aircraft carrier had been called back to Norfolk as a gesture to ensure the peace process in the fragile Middle East, and all was well in the world except for the fact that every country that could was trying to figure out the best way to deliver an atomic bomb without blowing themselves up while doing so.

When Darrell finally stepped back onto solid ground and onto the docks at Norfolk Naval Base, three weeks had passed, the battery was dead on the Mercury, and he still had no idea what he was going to say to Audrey when he drove home to Levittown.

Chapter 9

"We have invaded space with our rocket and for the first time, we have used space as a bridge between two points on the earth. We have proved rocket propulsion practicable for space travel. This 3rd day of October 1942 is the first of a new era for transportation . . . that of space travel."

— German General Walter Dornberger after the successful launch of a V-2 rocket from Germany during World War II

Everyone knew that the Russians had rockets. After all, the Germans had rockets during the war, and half of the Germans that made those rockets were captured by the Communists and had been working for the Soviet Russians ever since. What was going on behind the Iron Curtain was always a mystery, but atomic bombs and rockets were a foregone conclusion, and as anyone who was digging a bomb shelter knew, it was only a matter of time before Russian rockets were tipped with atomic bombs.

Detroit was into rockets in more ways than anyone knew, but rocket fins were certainly making appearances on most of the American cars rolling out of the assembly line. It was certain that rockets and spaceships were the way of the future because it was hard to find a movie without a rocket in the 1950s. There were Hollywood-manufactured rockets blasting off for Mars, Venus, and even Planet X, and onboard those shining silver tubes were glamorous movie stars ready to invade the imagination of anyone who was watching. There were little green men from outer space that came in flying saucers, and as the moviemakers in Hollywood decided, the only way to battle alien creatures from another planet was with rockets.

The U.S. Navy was into rockets, because the admirals at the pentagon felt that rockets were too sophisticated and too advanced for the U.S. Army. The generals in the Army believed that the Navy was wrong. The generals were tired of the Navy's boasting, and they wanted to prove that the Navy's rockets were not as great, perfect, and invincible as the admirals in the pentagon wanted everyone to believe.

The U.S. Navy's sparrow missile system, the Navy declared, could protect the entire U.S. coastline from Communist attack. Russian rockets and airplanes couldn't get past the precise and refined radar that guided the Navy's very special rockets, and this constantly broadcast rhetoric was beginning to make the generals in the Army furious. The Navy verses the Army was not anything new, but the generals felt they needed to prove the Navy wrong if only to offer proof that the U.S. boundaries were not safe from Russian or Red Chinese attack.

During World War II, when General George Patton was charging through France and Germany and leading his troops, Russian generals were fighting the Nazis and gaining ground from the east. All along the way, in both directions, there were fascinating German installations that were captured, and many of these hidden and heavily fortified manufacturing plants had obvious scientific value. There was a huge munitions factory under a mountain beside the Necker River and there were refineries that could process coal into diesel fuel and gasoline. There was another very well kept installation under several hundred feet of solid rock in the Harz Mountains, and that secret Nazi base was a real showstopper.

Two miles northwest of Nordhausen in central Germany, in an area known as *Mittlewerk*, the American Army "Timber Wolf" Infantry Division discovered two huge parallel tunnels running into two miles of chiseled-out stone. Inside the abandoned installation was a futuristic under-mountain rocket factory with several hundred rockets in various stages of completion. Many of the German ultramodern missiles were fully operational and ready for railway transport.

The German V-2 rocket was the world's first ballistic missile and was used to deliver one ton of high explosive into the heart of London, England, and Antwerp, Holland, at supersonic speed. Developed largely by an extremely youthful Dr. Wernher von Braun, the V-2 was an unmanned guided missile that could reach heights of fifty miles above the earth in about sixty seconds. During the war, the Nazis called the most modern rocket in the world a "Wonder weapon" because of the unprecedented technical achievement of von Braun's engineering.

The V-2 was a liquid-fueled rocket using a combination of methyl alcohol and liquid oxygen. Pure hydrogen peroxide and calcium permanganate were used as an instantly ignitable steam source, and when super heated steam powered an extremely powerful turbo fuel pump, the German V-2 was the first manmade creation to ever pass through the edge of space.

After the American Army "Timberwolves" found the Nordhausen installation and the grisly Camp Dora concentration camp nearby, a shroud of secrecy covered all aspects of the German rocket program. A clandestine and undercover search was instantly launched for the creators of the rockets and all of the technical documents that were obviously missing. Shortly after the Nazi surrender, many of the completed rockets and all of the German technicians that could be found were shipped back to America in what later became known as "Operation Paperclip."

The German rockets that arrived in America were at once made available to the U.S. Army and Navy for experimental research and were stockpiled in military warehouses all along the eastern seaboard.

When a frustrated and concerned U.S. Army general called the Sperry Corporation and asked to speak to an expert on guidance systems, Sperry called one of their brightest new technicians. His name was Darrell Loan. After a meeting was arranged, a group of Army staff cars arrived in Long Island.

"Loan," the general began, "I want you to defeat the Navy's sparrow missile system. Sperry developed the sparrow missiles and now they tell me you're the right man for the job." The general frowned and then stared hard at the young civilian with the blue eyes and the sandy hair that was standing before him.

When his gaze shifted to his attending officers, the general continued, "All of the admirals at the pentagon are walking around and acting as if the U.S. boundaries are impervious to an airborne attack," the senior officer shook his head. "But I don't believe it for a minute and I want you to prove them wrong."

"Of course General, I'll try," Darrell responded. "But I'll need to have all the technical information on the sparrow system. Everything you can get from the Navy, and all the original specs from Sperry."

"We're already ahead of you." The general motioned to an aid that produced an army-issue briefcase.

"Locked away in this not-to-leave-this-room case — is every detail about the Navy's little rockets: all the specs, the guidance information, and the targeting radar that is linked to the system. Of course, all of this is completely confidential. I'll look forward to hearing from you. Call me anytime, but wait until you have some answers."

After the meeting, Darrell poured over all the Navy specs and pages of information. The following day he called the general who was back in Washington.

"General, I believe I have your answer," Darrell explained over the phone. "We need to have another meeting."

No more could be said over the telephone wires. The threat of Russian espionage was advanced and very real.

Chapter 10

"The atomic bomb has made the prospect of future war unendurable. It has led us up the last few steps to the mountain pass and beyond there is a different country."

— Robert J. Oppenheimer

At a secure conference room hazy with cigarette smoke, and at a table lined with Army officers, ashtrays, and coffee cups, the young troubleshooter from Sperry and the only man not in uniform began to share his idea. Everyone was smoking.

"Can we get our hands on one of those V-2 German rockets that were captured at the end of the war?" Darrell asked. "One that is functional?"

After a quick turning of heads and a huddled muffle of conversation, the general nodded. "Yes we can," he confirmed. "I have just been advised we have about eight in reserve that have never been used."

Darrell grinned. "And can we have some technicians that are familiar with how to fuel and launch one of those rockets?"

"We can do that." The general lit a fresh cigarette. "No problem there. Why do you need that old German ordnance?"

"Because, the Navy's sparrow missile system relies on a radar that cannot detect anything below five hundred feet. Anything coming in below the sparrow radar will be undetected, and I can program the gyros and the guidance of the

V-2 to cruise below five hundred feet. The V-2 will fly horizontally, under the radar, until it reaches a target. I'm sure it will work."

With a burst of enthusiasm, the Army officers all began talking at once. After a few moments, the general called for quiet. "All right son, you've got yourself a rocket."

Chapter 11

The V-2 rocket program was the most expensive project in all of Germany during the Second World War.

With the Navy placed on a mysterious alert, and a sparrow missile system positioned on active standby, a group of steadfast Army engineers began carefully fueling a thirteen-year-old German rocket. There were dangerous, volatile, and extremely unstable ingredients required to make the war relic fly and with methyl alcohol, liquid oxygen, pure hydrogen peroxide, and calcium permanganate on the engineer's fueling list, even the boldest of the Army technicians was cautious, nervous, and worried.

The liquid oxygen was three hundred degrees below zero and two of the technicians already had frostbite from accidentally touching metal couplings coming out of a fueling truck. As the liquid oxygen vented as a rolling fog pouring down the outside of the rocket, a thin sheet of ice formed on the captured trophy from the secret Nazi base. The pure hydrogen peroxide had been loaded with the utmost care, as even one fly-spec of organic matter in the glass-lined fuel tank would have created an instant explosion. Filling the adjoining container with calcium permanganate was just as dangerous, because when the peroxide and the calcium touched, there was a hypergolic explosion that created massive amounts of super-heated steam. All of the extremely dangerous components were necessary, the Army engineers knew, but what everyone was wondering was whether the German engineers that had been trained in the fueling procedure thirteen years earlier had important safety precautions the American Army didn't know anything about.

With the liquid oxygen nozzle the last to be removed, one of Wernher von Braun's creations sat beside a service gantry on a concrete launching pad and pointed skyward. To the army engineers and officers that were watching, the aging German rocket looked ominous, dangerous, and like an explosion waiting to happen. With the liquid oxygen chilling the metal, the slightly scraped and dented V-2 creaked, popped, and hissed, and sent out intimidating signals as Darrell Loan, dressed in civilian clothes, completed a quick and final check of the guidance control package in the nose of the rocket. As he was climbing down from the service gantry scaffolding, two olive green sedans labeled with white Army stars were approaching. The Atlantic Ocean was in view, gray with whitecaps, a strong breeze carried the scent of sea salt, and all along the coastline, waves of sandy dunes marched into the distance. There were no houses and no buildings, only a concrete launching pad and the beginning of a new era.

Chapter 12

"Courage is being scared to death . . . but saddling up anyway."

— John Wayne

With the arrival of the two Army staff cars, the general walked forward with his adjutants and everyone gathered behind a makeshift wall of fortified sandbags. The sandbag wall was three bags thick, forty feet long, and ten feet high. The Army was taking no chances.

"You really think that smoldering hunk of junk can defeat the Navy's sparrows?" Every ironed crease on the general's uniform was perfect as he stood on an observation platform. He was behind the sandbags and looking over the protective barrier toward the camouflage-painted rocket. His question was directed to the only person not in uniform. All of the officers, most of the fueling technicians, and even the enlisted men that drove the fueling trucks were watching from the elevated platform and the safe side of the wall.

"Yes sir I do," Darrell's response was confident. "As long as she flies the way she was designed we should have no problem—as long as everything still works after twelve years in a warehouse."

"It might have looked impressive to Hitler," the general frowned and his officers looked skeptical, "but right now I just can't see it. That thing looks like an over-engineered nightmare ready to go bad."

Before anyone could comment, the chief technician in charge of the fueling procedure stepped up to the officers behind the sandbags. "General," the soldier began and saluted. "The V-2 is ready to go, and I think we should light this candle as soon as possible. There's nothing in the translated manuals that mentions launching right away, but it seems like there's a lot missing out of those old instructions. I just have a feeling this thing is ready to start without us."

"All right Loan," the General nodded. "Let's get this show on the road."

With everyone watching carefully, from just over the top of the sandbag wall, a quick countdown procedure began, a servomechanism opened the hydrogen peroxide valve, and a sparking device began the ignition sequence at the base of the rocket engine. When the peroxide touched the calcium, everyone could hear the fuel pump turbine begin to accelerate and whine, and when the first of the alcohol and the liquid oxygen ignited, an explosion of thunder shook the ground as the German rocket rapidly began to rise above a column of superheated flame.

When photographs are taken during a rocket launch it is very important that the photographer understands that there are only seconds available before the rocket rises above the service gantry and is no longer present for photographs. In the Second World War, when the Germans were launching the first ballistic missiles, the V-2 rockets were fifty miles high in sixty seconds and traveling over fifteen thousand miles per hour.

From the remote Atlantic beach, in less than two minutes, the war relic of the Third Reich operated perfectly and its flight was finished. With the general, Darrell Loan, and all of the Army personal watching, the V-2 was indeed a spectacular sight as it rose above a fiery tail to a height of three hundred feet and rolled over to fly horizontally. It then disappeared toward a target that was under the protection of the Navy's sparrow missile system.

The Navy never saw the rocket coming. The sparrow radar could not detect the V-2 coming in at near the speed of sound, below five hundred feet, and after the Secretary of the Navy heard the news, he launched an investigation to find out who was directly responsible. He was furious.

Chapter 13

"Everyone has his day and some days last longer than others."

— *Winston Churchill*

At the Pentagon in a room full of Admirals and Navy brass, the details of the unbelievable defeat of the Navy's sparrows began to rise to the surface. All of the naval officers were seated at a conference table lined with coffee cups, ashtrays, and a printed report labeled: "Top Secret." Everyone was smoking.

When the Secretary of the Navy entered the room all hands were on the deck standing and saluting, but the Navy was ready to get underway so everyone settled down to listen.

"The Sparrow Missile System was defeated last week by a civilian from the Sperry Corporation in Long Island." The naval officer reporting was an immaculately dressed staff captain who had always worked at the pentagon and was always awash in paperwork. He had never been to sea.

"We have determined, and verified, through our usual channels that one of the 'Operation Paperclip' rockets was programmed to fly under the sparrow radar."

"Under the radar . . ." the Secretary of the Navy leaned back in his chair as he repeated the key words. When his fingers formed a steeple, the reporting staff officer knew to remain silent until called upon.

"What we have here is an entirely new weapons system," the voice behind the arched fingers began. "Conventional rockets that are not so conventional . . . with a guidance system that can fly them below enemy radar. There could be no better way to launch such a weapon than from a ship or perhaps even a submarine."

The chief of the U.S. Navy had a faraway look in his eyes. "I'll bet the Army doesn't even know what they have stumbled onto. This is absolutely brilliant."

Every man in the room was surprised. The Chief Naval Officer for the United States was no longer angry. He was impressed and inspired.

"What's the name of this civilian that defeated my sparrows with an old German rocket?" The Secretary of the Navy was now focused on the reporting staff captain.

The officer reading the report was ready. "His name, Mr. Secretary, is Darrell Loan."

When the naval branch of the Pentagon called and asked to speak to Sperry's chief troubleshooter and the individual responsible for the V-2–Sparrow incident, it was only a matter of minutes before the Secretary of the Navy was on the phone. Sperry knew the call was coming, and everyone was ready.

"Mr. Loan," the naval chief began carefully. He knew about the Russians and the phones. "That thing you did for the Army last week . . . I would like to ask that you don't do that again. Do we understand each other?

"Yes sir, I believe we do," was the only answer possible.

"Good, but if something like that should happen in the future, I would ask that it be sent a little higher, so our people could have a good look at it. I would really appreciate it." The naval commander's voice shifted gears. "Call me if you ever need anything. Oh, by the way, I want to thank you for that work you did for us awhile back — on that occasion you had to spend time away from your home and your family. Don't forget, if you ever need *anything*, just call. Goodbye."

When Darrell hung up the phone, he was smiling.

Chapter 14

"Second star on the right . . . and straight on till morning."

— Peter Pan

Audrey never knew what Darrell was working on — because he was not allowed to tell her — but she was tired of New York and all the men in suits that may or may not have been Russian spies or FBI men following Russian spies, and she was worried that Darrell might disappear again, and this time forever.

The edgy New Yorkers were abrasive, especially for folks from the Midwest, and ever since the day that Darrell disappeared for three whole weeks with only a cryptic phone call, for Audrey, the comfort level at the rented home in Levittown was tense at best.

She knew his work was important, but she could not help but wonder whether farming back in Iowa would have been a much better life. When the man from Chrysler called, Audrey felt her spirits lift and her outlook brighten. His voice was friendly on the phone, mainly because he didn't have that dreadful New York accent, and probably because he did have manners.

"Mrs. Loan, my name is Bill Hinkle, and I'm with the Chrysler Corporation. You know, Chrysler Automobiles," the easygoing voice explained. "I'm calling long distance from Detroit."

"Yes, Mr. Hinkle, what can I do for you?" Audrey's response was guarded even with a friendly voice. You had to be suspicious in New York, especially if your husband worked on things that you weren't allowed to know any-thing about.

"Mrs. Loan, I'll just get right to the point," the cheerful remarks from the Midwest continued. "We want your hus-band to come out and work for us here at Chrysler—out here in Michigan. I'm calling you because every time I try to reach your husband at Sperry, I can never get him on the phone." Bill Hinkle laughed. "The trouble is I have a feel-ing that Sperry doesn't want me to meet your husband. I'm thinking that they know I want to offer him a new job. We think you folks would be really comfortable out here in De-troit."

Suddenly, Audrey's mind was racing. She loved the idea. Anything would be better than New York. Michigan was much closer to Iowa and home, and she hadn't heard any-one say *folks* in a long time. This was exciting. Darrell would be making automobiles and that was something Russian spies or the FBI wouldn't care anything about.

"Mr. Hinkle," Audrey couldn't keep the happiness out of her voice, "I think that would be a wonderful idea."

"Well good!" Bill Hinkle laughed again and Audrey could feel the warmth. "I could fly out to New York next week. Do you think you could convince your husband into having dinner with me? I could present a proposal that I believe you *both* might find very attractive."

"Oh yes, I think I could talk him into that." As soon as she said the words, Audrey crossed her fingers for luck. In an instant, she imagined Darrell called away on some mys-terious project and forced to miss the meeting. She was so very tired of all the secrets at Sperry.

"Let me tell you something," the nice Midwest voice was speaking again. "You talk to your husband, find out when would be a good time to meet, and tell him I promise to buy him the best steak dinner in New York City if he'll just listen to what I have to say."

"Okay, I'll try." Audrey tried to keep her voice calm and friendly, and her desperation hidden.

"Do you have a pencil? I'll give you my phone number. I want you to call me collect."

Quickly, Audrey scrambled into her purse and found a pen. She wrote the Chrysler number on a pad by the phone and promised to call as soon as she could.

"Please let me know right away," Bill Hinkle's voice was even more cheerful. "I'll look forward to it!"

After she said goodbye and hung up the phone, Audrey went shopping. She now had a plan to get out of New York.

Chapter 15

"Both the man of science and the man of action live always at the edge of mystery . . . surrounded by it."

— Robert J. Oppenheimer

Work at Sperry was going great and Darrell really liked his job. There were plenty of exciting challenges despite the unpredictable hours, the remote assignments, and the confidential meetings, and the people at Sperry were kind and appreciative, and the money was good.

Over dinner one evening, when Audrey carefully brought up the phone call from Detroit and the conversation with Bill Hinkle, Darrell was skeptical.

"Audrey, you know I like working at Sperry, and my boss is the greatest. We're making new discoveries almost every day. The work is very exciting."

Audrey let that one go. She knew she was treading into unknown territory, but she also knew what she wanted.

"But Darrell," Audrey pleaded, "Don't you think it would better in Detroit? We would be closer to home. After all, we have our family to think about *and* the future. I know you don't like the New Yorkers any more than I do, and besides, I really don't have any friends out here. I just can't take it if you disappear again."

Darrell was just finishing a big plate of pork roast with roasted onions and potatoes, and Audrey was ready to bring out dessert — strawberry shortcake. The meal and the evening had been carefully arranged. Audrey knew just what to do.

"Would you like some more wine or coffee? By the way, how do you like my dress?" She knew she looked good in yellow, and earlier in the day, Audrey had been to the hairdresser.

Darrell grinned. "All right Audrey, you win. Go ahead and call the man from Chrysler. I can meet him next week."

Chapter 16

"The best way to predict the future is to create it."

— Abraham Lincoln

When Bill Hinkle arrived in New York City, he knew he was a man on a mission. The Big Apple was booming as only New York can, and after checking into the Waldorf Astoria and making dinner reservations, he had time to reflect on his instructions from Chrysler.

"Bill," the special projects director had begun, "we are beginning to fill these specialty jobs nicely, but I can't tell you how important it is to find the right people for our new Jupiter project. This could be the greatest thing for Chrysler ever. Absolutely the greatest! There's no telling where this thing could go, and I really want that young man from Sperry."

"Why is he so important?" Bill remembered asking the question in the big executive office overlooking the Detroit skyline, but when the answer came, it was a little unnerving.

"Because the government says we need him, the Army insists we have him, and because he's the best troubleshooter Sperry has ever seen. This guy can think on his feet and he can do it fast. He's a real problem solver and works great under pressure. That's why we need him, and that's why you've got to convince him that we're better than Sperry. Bill, this is really important. Offer him anything he wants."

A week earlier, Bill Hinkle had received the call from Audrey Loan in Long Island. She explained that Darrell had reluctantly agreed to meet for dinner, but that was it, just dinner—no commitments.

With the elevator full, and the Waldorf filled with bustling businessmen and tourists, Bill Hinkle took in the perfume of a beautiful blonde next to him and waited for the elevator boy to open the door.

When the elevator opened on the ground floor, the lobby was packed. Gray suits were the rage in New York City, and almost every man was dressed in a variation of the monotone, but *everyone* had the customary white handkerchief tucked into a left breast pocket. Women were everywhere clattering across the marble floor wearing high heels with white gloves and purses, and they were wearing hats of every color, fashion, and size imaginable. If the Russians were going to blow up the world, the women in New York City were going to look good when it happened.

Bill approached the reception desk and looked around. He was on time, ready for dinner, but also a little worried. All he had was the memory of a grainy photograph of the young man from Sperry, and he couldn't really remember what he looked like. What Bill did notice, however, was that his suit was blue and that almost all of the other men were wearing the loose-fitting gray suits. Everyone, Bill decided, but the kid that was watching him. He was also wearing a blue suit. Suddenly the kid smiled, and Bill knew just as suddenly that it was Darrell Loan.

Bill crossed the lobby to where his job was waiting. He returned the smile and offered his hand. "I was worried that I wouldn't be able to find you, or that Sperry would have sent you off on the road."

The kid from Sperry grinned as he shook hands and Bill thought: *He doesn't look old enough to rate this kind of money.*

"My wife mentioned that you wanted to buy me a steak dinner. When I heard it was here at the Waldorf, I knew I couldn't miss the meeting."

"That's great Darrell, just great," Bill Hinkle said. "Can I call you Darrell?"

"Sure, everybody else does."

"Are you hungry—ready for dinner?"

"You bet. I could eat a horse."

Walking across the lobby Hinkle suddenly felt uncomfortable and somehow nervous. This kid had all the confidence in the world—you could just feel it. It was not just that he was taller. He was tall—easily over six feet—and thin, but he didn't look like a typical engineer. He looked like a fresh-faced actor straight out of Hollywood. His complexion was clear, he had sandy hair and blue eyes, but there was something more—something beneath the surface—something that couldn't be measured. His walk was like something out of a cowboy rodeo.

After they were seated, a waitress came and took drink orders. She obviously liked the kid and focused on him. Bill felt like he was watching a movie.

"What can I do for *you* slim?" After she asked, the pretty brunette smiled, and Bill thought: *Is this how cocktail service is offered at the Waldorf?*

Darrell's smile was spontaneous. "I'll have a double martini with extra olives."

Hinkle interrupted the eye contact, "I'll have the same."

When the waitress was gone with a wink, the man from Chrysler knew he had better get down to business.

"You're a hard young man to get a hold of," Bill Hinkle offered a cigarette and a light. Across the dining room, everyone was coming in for cocktails and dinner. New York City was hungry.

"Sperry keeps me hopping." Darrell was looking around the restaurant and then out to the lobby. Beyond the main entrance the city streets were packed. New York City was busy.

Hinkle was hoping for a little more opening conversation, but the kid from Iowa seemed preoccupied and noncommittal.

"Darrell, I'll just get right to the point. I know your wife must have told you—have you ever thought about changing jobs—about coming to work for us over at Chrysler?"

"No," Darrell said. "I like Sperry and I like the work."

"You know, ever since Sputnik, things have changed. There's something coming that's going to be bigger than anything that's ever happened. Chrysler is going to be part of that—a very big part. We have the government contract for the Army Ballistic Missile Agency. When Sputnik went up, that contract just got a lot bigger, and a lot more urgent."

Before any comment or answer was possible, the pretty brunette returned with drinks, and a waiter came and took orders for dinner. Darrell ordered the big T-bone and Bill Hinkle the same.

"Thanks for dinner," Darrell offered when the waiter was away and a moving part of the dinning room.

"Anytime," Bill Hinkle raised his martini. "Here's to a new future and getting ahead of the Russians!"

After the toast, Hinkle said quietly, "Can I ask you a question?"

"Sure, and I'll answer if I can." Darrell sipped his drink.

"What kind of money are you making over at Sperry?"

Darrell grinned, "None of your business."

"I didn't mean to be so direct, or to pry, but I have to tell you: This job at the new Chrysler missile division will have a top government priority. Do you know what that means? America can't take the Russians being first in space. If we fall behind, and can't keep up, that just means the commie hardware is better than ours, and a war between the free world and the Soviets is all but over. Darrell, it's as simple as that. We have to win the race for space or everyone will know we can't win a war against the Russians and their hardware.

"You're not telling me anything that I don't already know, but my work at Sperry is important. We're helping our cause a lot . . . believe me." Darrell leveled his gaze and the man from Chrysler thought: *I've never seen a kid this young so confident.*

Before the flirting waitress could return tableside, Hinkle motioned for another round of drinks and hoped that would help.

"Darrell, I know that you have a security clearance a lot higher than mine, but there's something going on that I'll bet you *don't* know." Hinkle suddenly knew this was the right approach. There was a notable change. The kid looked interested. If there was something he didn't know, he wanted to find out.

"Eisenhower is getting nervous," the Chrysler man confided, "and from what I've been told, he has a good reason. Right now, the Navy is in charge of sending up our version of Sputnik. They have a prototype rocket they have named Vanguard, but the president wants a backup plan. He's an Army man, and he wants the Army to have a rocket that works if the Navy can't pull through. He doesn't have a choice. Everybody knows we have to get a satellite into space and we have to do it fast."

Darrell remained silent, his eyes steady. Clearly, he wanted to hear more.

"There are some experts," Bill Hinkle leaned forward, "that are confident the Navy is going to fail."

"What kind of experts?" Darrell munched an olive.

"Let's just say, Chrysler is doing what we do best." Bill Hinkle's voice was now a whisper. "We have a design team to make the body, and another team to make the engine. The boys on our engine team are the experts. Rockets need engines that work."

After drinks, dinner, and dessert, the man from Detroit ordered coffee. He needed it.

When the waiter was away, Darrell said what had been on his mind. "What makes your Army experts better than the Navy?"

"Experience." Hinkle tasted the coffee. "Our experts have done this before, years before. There's no better rocketry team on the planet."

"Are you telling me that you have the Germans?" Darrell for the first time looked surprised. "The Germans from the war?" he said. "The same Germans that built the Nazi V-2?"

Bill Hinkle leveled his gaze, and delivered his invitation to Detroit. "I've said all I can unless you take the job, but I *will* tell you this: I am offering triple the amount you are making at

Sperry. Three times as much. You show me what you're making, and we'll pay three times as much. What do you say?"

"I'll think about it." Darrell nodded. "Thanks again for dinner."

"My pleasure, call me when you've finished thinking." It was Bill Hinkle's turn to nod. "Your wife has my number. Call me collect."

Chapter 17

"Tact is the art of making a point without making an enemy."

— *Isaac Newton*

When Darrell came home, Audrey was waiting. It was late, almost midnight, but Audrey was dressed and her hair was perfect. She was waiting by the door.

After a quick peck on the cheek she smiled, "How about a nightcap?"

Darrell's smile was patient. "Sure."

When the drinks were made, Audrey settled in on the sofa. "How did it go?" she asked. "Was he nice?"

"He was very nice and dinner was great. We both had steaks. The Waldorf is incredible."

Audrey patiently sipped her drink. She was determined to beat Darrell at his perfected persona of composure and nonchalance. "The paper came today," she began. "There's a great advertisement for Chrysler. The new cars really look good — much better than Chevrolet. I really like the look."

Darrell nodded, but remained silent.

"I really think that Chrysler would be good." Audrey tried to sound as if this wasn't practiced. "I'll bet if we moved to Detroit, we could get a great deal on a new Chrysler. Just think, you could be home every night for dinner and drive a new car.

"I already drive a new car, and I like the Mercury. I think the Chryslers look boxy."

Audrey smoothed her pleated skirt, "But the Mercury is not ours. It belongs to Sperry. Don't you think —?"

"Audrey," Darrell interrupted softly. "Your man from Chrysler, he offered three times the money. Triple what Sperry pays."

"What?"

"You heard me," Darrell couldn't stop the smile.

"Then are we going to do it—are we going to move?" Audrey was on her feet, her excitement overflowing. "Oh Darrell, I just love the idea of you building cars for Chrysler and a nice stable job without all the secret stuff at Sperry!"

"Audrey," Darrell shook his head. "I told the man I would think about it. I didn't say I would do it. I still have to talk to my folks at Sperry."

Chapter 18

"There are people who make things happen . . . there are people who watch things happen . . . and there are people who wonder what happened. To be successful, you need to be a person that makes things happen."

— James A. Lovell

The meeting with Bill Hinkle was on Saturday night, and on Monday morning, Darrell went to work as always. He did, however, drop by his boss's office, slip past the secretaries with a smile, and take a seat in front of Alan Misner. Alan had been Darrell's first and favorite boss at Sperry, and he had always been fair, appreciative, and friendly.

After answering questions about Audrey, the home in Levittown, and a new engine analyzer that was an ongoing problem, Darrell told his boss about the meeting with Bill Hinkle from Chrysler.

"Darrell," Alan began. "I should have seen it coming. Chrysler and Sperry have worked together for years. It only figures that they know what we're doing and who to keep an eye on. But *I know* what's going on at Chrysler, just like they know about us."

Alan punched the intercom button on his desk. "Joyce, hold my calls, and send in some coffee."

When the coffee came, Misner began, "Until now the president has backed the Navy's Vanguard booster to get a satellite into orbit, because he has largely ignored anything that Wernher von Braun and his Germans have come up with. If the Germans submit a plan, Eisenhower drops it in the trashcan. Ike's old school and he can't help but remember his fight against the Nazis. The Navy promised they could get Vanguard up and off the ground for twelve million, but now that figure is through the roof, and they are up to one hundred million and rising."

Alan Misner shook his head. "That's a figure that Congress just can't take. Luckily, for von Braun, an Army general named Medaris has been backing him all along. He's the man in charge of the Army Ballistic Missile Agency. Quietly and on the sidelines, Medaris has backed von Braun and his Germans, and now they have a rocket named 'Jupiter' that will hopefully lift an American Sputnik."

Alan smiled and opened his hands. "That's why they want you. They have a rocket ready to go—at least on the drawing board—but they must have troubleshooters in every field if they want to produce what they've promised."

Darrell wasn't surprised that Sperry knew all the details about the Chrysler missile program but the next part was unexpected.

"There *is* something else," Misner paused, "that I think you should know. After the war, the Russians captured as many of the Germans as they could that were involved with Nazi rockets. Everyone says that von Braun is the best, but there were many good technicians that were forced to go behind the Iron Curtain and work for the Communists. They were captured to do just what von Braun is doing here, but now the Russians think they know all there is to know about rocketry, and they are sending the original Germans back over the border to West Germany."

Darrell leaned forward, "That seems crazy," he said, "or very stupid. Why would they do that?"

"Of course it's crazy." Misner stood up and began to pace. "As soon as the German engineers made it out of Russia, they began searching for their old boss von Braun. It wasn't hard for them to make contact, and the Defense Intelligence Agency soon began to round them up. But now the Russian-captured Germans are being sent over here. Along with everything they know."

"But if these guys know how the Russian rockets are made," Darrell's mind was racing, "we could learn a lot and make all kinds of breakthroughs. After all, the Russians seem to be leading the way. If we can learn what those German technicians know, we would save a lot of time and money."

Alan Misner frowned. "Yes, but what if it's one of the oldest tricks in the book? A Trojan Horse filled with secret surprises, sabotage, or even a spy or two to report back to Mother Russia."

"That *is* something I should know. Something we all should know," Darrell looked thoughtful, and then it was his turn to frown, "and a lot to think about." He offered, "What do you think I should do? Should I take the job?"

Alan shook his head and smiled, "Darrell, you have no choice—" he said, "for three times the money? You would be crazy not to take it. Besides, I think we need someone at Chrysler that knows how to look out for trouble. One thing is certain: if you take that job in Detroit, you're certainly going to earn the paycheck."

Chapter 19

"Space isn't remote at all. It's only an hour's drive away . . . if your car could go straight upwards."

— Fred Hoyle

December 6, 1957

Audrey was standing in the living room with her hands on her hips. Darrell was in front of the television changing channels.

"I really don't want to do any decorating," Audrey said. "Not if we are going to be in Detroit for Christmas. Won't that be exciting — a new home and a new job for Christmas?"

Darrell stood up and backed away from the television. On the black-and-white screen, a rocket was standing on a launching pad. Clouds of venting vapor were rolling down the rocket's exterior as it stood next to a service gantry tower. Spotlights were shining on the sleek white missile with a black upper stage, and there were palm trees visible in the Florida background. It was a live shot with television crews and newscasters anxious and eager for the upcoming launch.

Audrey stepped closer to the television and turned up the volume. ". . . *The United States* . . ." the newscaster began, *"has long awaited the launch of Americas' first attempt at space . . . and tonight, in the balmy Florida tropics . . . the mighty Vanguard Booster stands tall and proud as Uncle Sam rises to the occasion and joins the race for space."*

Audrey stood back with her hands on her hips. She then untied her apron and sat down on the sofa. "Wow!" she said as she leaned forward. "That's quite a rocket."

Darrell grinned. "That's the Navy's rocket," he said. "They sure are under a lot of pressure to make this work."

Audrey was focused on the television. "I just keep thinking about that poor dog the Russians sent into space. They let her run out of air and she died up there. It just doesn't seem right—that dead dog, up in space, and circling over our heads every hour."

"Actually it takes about and hour and a half," Darrell offered, "for the Russian satellite to complete an orbit."

"Shush!" Audrey said. "Listen!" The newscaster was wearing a black tie and a white shirt with the shirtsleeves rolled up to the elbows. His background was a stand of palm trees and, in the distance, the smoldering Vanguard rocket.

"*. . . Dr. Hagen with the Naval Research Laboratory has been working very hard with his team to ensure the Vanguard booster meets the Soviet challenge to launch a peaceful effort for the exploration of outer space.*"

"How exciting," Audrey was watching the black-and-white screen intently. "How much longer before it goes up?"

"Any minute, but there's a lot of factors," Darrell explained. "The weather looks good. Temperature seems right, but everything has to work perfectly, and in a perfect sequence. A lot can go wrong. Rockets can be tricky."

"How do you know that?" Audrey looked over, "How could you possibly know *that*?"

Darrell shrugged as the television reporter continued, "*. . . We seem to have some activity. The last connection to the service gantry is being removed . . .*" Suddenly the image of the reporter was replaced with the full view of the rocket and the chilled liquid oxygen vapor venting and pouring down the sides.

"*. . . Ladies and gentlemen, we should have the launch of the Vanguard booster at any moment . . . and America joining the long-awaited race for space.*"

With Audrey and Darrell watching along with most of America, Dr. Hagen's Vanguard booster began an ignition sequence with a bright burst of flame at the rocket's base. Very slowly, the very tall and pencil-like missile began to rise from the launching pad.

"... *This is it ladies and gentlemen* ... *with the liftoff of the Vanguard booster and the satellite TV-1 America's venture into —*"

All commentary stopped as the Vanguard booster hovered at about five feet above the earth and then began to settle backward into its own exhaust flame. When the rocket engine failed to produce enough thrust to lift the missile, and the rocket began to falter and fall, the entire launching pad and the service gantry tower were instantly consumed in a huge fireball as the pressurized liquid fuels detonated. The violence of the Vanguard explosion was beyond tremendous.

"... *What a terrible failure ladies and gentlemen* ... *What a terrible tragedy* ..." the newscaster had regained his voice, but the words were shaky and searching. "... *The nosecone of the rocket holding the beeping satellite has been blown clear of the explosion and is apparently still beeping* ..."

"Oh Darrell, how terrible," Audrey was on her feet. "It's almost like you knew that was going to happen! Did you know? Is this part of what you are doing at Sperry?"

"Audrey," Darrell sighed, "I guess I had better tell you right now that my new job at Chrysler is not going to be making cars that look boxy."

Chapter 20

"Opportunity is missed by most people because it is dressed in overalls and looks like work."

— Thomas Edison

Three days after the steak dinner at the Waldorf and the next day after the Vanguard booster exploded on the launching pad at Cape Canaveral, Darrell called Bill Hinkle at Chrysler.

"Bill can you hear me?" Darrell spoke over the static-filled long distance line. "I'll be happy to take the job. When do you want me to start?"

"We want you right away." Even with the long distance connection, it was obvious that the man from Chrysler was delighted with the news. "I know that you'll really like Detroit. Don't worry about moving. We'll send movers and get you settled in quickly. You'll be living here in Warren, Michigan — that's Chrysler world headquarters." Static crackled over the phone, and Bill Hinkle said, "I'll meet you at the airport. Can you fly out tomorrow?"

With New York decorating for Christmas, Audrey and Darrell took a cab to Idyllwild airport and flew to Detroit. The move was easier than anyone could have imagined. Chrysler took care of everything, including first class airline tickets from Trans World Airlines. Onboard the airplane, the men wore suits and the women hats and gloves. Iced shrimp cocktail was served in crystal serving dishes and dinner was roast duck on TWA signature china. TWA also served the best coffee ever from Columbia, South America.

When the Howard-Hughes-designed Constellation airliner touched down in Detroit, it was snowing. As promised, Bill Hinkle was waiting at the airport. He met Audrey and Darrell inside the terminal and led the way out into the frosty air where a Detroit cop was waiting beside a brand new Chrysler. The new car was obviously parked illegally, right up and onto the curb. The cop smiled, nodded, and shook hands with Hinkle. He then helped with the suitcases. After the luggage was in the trunk, Bill drove through the evening snowflakes down Telegraph Road. No one said anything about the cop. Bing Crosby was singing *White Christmas* on the radio.

"Darrell, I just love it," Audrey said when they arrived outside their new home in Michigan. "It's perfect, and I love the snow. The snow in New York is dirty."

"The movers should be here by the day after tomorrow," Hinkle said as he stomped the snow from his feet on a big welcome mat.

Light flurries had been falling since the plane landed at sunset, and now as the Chrysler man found the key and paused at the door, Michigan was a Winter Wonderland. The gently falling snowflakes were big and the temperature pleasant. Red and green Christmas lights were glowing from under the snow on a neighbor's decorated spruce.

After they were inside, Bill Hinkle said, "The heat is on, and we've got groceries in the refrigerator and in the cupboard. There are clean sheets on the bed and fresh towels in the bathroom. I think we've thought of everything, but if there's anything you need, just give me a call."

Bill smiled and Audrey hugged him. The living room was finished with knotty pine and there was a big brick fireplace with firewood ready on the grate. The house smelled musty, but it was much bigger and nicer than the rented home in Levittown.

Audrey took off her hat and coat and pulled off her gloves. "I repeat," she said with a smile, "I just love it."

Darrell received a peck on the cheek, and Audrey was off to the kitchen and then to the other rooms. "I love the kitchen . . ." her words trailed off.

"Thanks Bill," Darrell said. "You've really gone over-board."

"Don't mention it." Hinkle waved off the praise. "Besides, my boss tells me you're going to be worth it."

Darrell grinned. "I hope so," he said. "When do I start?"

It was Bill Hinkle's turn to smile, "How about tomorrow morning — at seven sharp? I'll drive you out to the plant."

"That's fine," Darrell was looking around. "Audrey will have me working here if I don't go to work for you."

"Two bathrooms!" Audrey's voice from across the house was delighted. "I can't believe it!"

"The fireplace ready to go?" Darrell asked as he walked over to the brickwork.

"Absolutely," Bill said as he moved toward the kitchen. "How about a drink to take off the chill?"

"Sure," Darrell was rolling up newspaper next to the hearth. There was newspaper, kindling, and fireplace matches. After a moment, the wood caught, and the fire began to crackle.

When Bill returned, he had two tumblers. "Is scotch okay?"

"Perfect for winter," Darrell said, and then he raised his voice to carry across the house. "Audrey, you want a scotch?"

"You bet!"

Bill said, "I'll get it Darrell. You relax."

With the fire crackling nicely, Darrell sipped his drink and walked over to a large bay window overlooking the snow-covered front lawn. The snow was still coming down by the streetlights and visible from the Christmas lights on the tree next door.

Bill crossed the room after placing Audrey's drink on the coffee table.

"You were right," Darrell said. "I like it here just fine and I haven't even seen the rest of the house. There's something about the feel of it. I'm glad to be back in the Midwest."

Bill tasted his drink. "Wait till you see where you'll be working! I guarantee it will be like nothing you have ever seen or believed possible. Imagine a hanger the size of four football fields under a roof with overhead cranes that can lift almost anything. We have a metal fabricator that can roll Reynolds aluminum into a perfect cylinder for the Redstone outer skin, and we have X-ray machines to travel down the welds and make sure every seam is flawless. Darrell, you're going to be impressed. This new Jupiter-Redstone project is going to be really big. The biggest thing ever."

When Audrey came into the room, she was glowing, and not just from the firelight. She was very happy. After picking up her drink, she took a sip and joined the men by the window. "It's beautiful out there," she said. "And in here, too. The firelight is wonderful." She put her arm around Darrell's waist and gave him a squeeze. She took another sip. "What are you two boys talking about?"

"We're talking about building rockets—" Darrell said and then he smiled, "about rockets that will work."

Chapter 21

"The creative conquest of space will serve as a wonderful substitute for war."

— James S. McDonnell

The first day at the fine family of Chrysler missiles was beyond incredible in many more ways than one. Darrell thought privately that if the quality of work that was clearly evident at the rockery division was applied to making the boxy cars, Chrysler automobiles would be the best in the world.

Bill met Darrell at the new house at seven sharp and they drove out through the snow to the Chrysler Industrial Reserve Aircraft Plant at the sixteen-mile road north of Detroit. After Hinkle opened his wallet and showed his identification, he presented Darrell and his Crypto security credentials to the guards at the security checkpoint. When the security procedures were complete, a briefing tour began in the big Redstone hanger.

The plant was just as Hinkle described, but even more impressive. All of the technicians were wearing white uniform coveralls but no one was dirty. Each section of the massive hanger was dedicated to specific areas of the Redstone construction, and there were several of the seventy-foot rockets in various stages of completion. The incredible installation was filled with the sights and sounds of the production on a rigid schedule. Even though the snowfall had continued overnight and most of the cars were chugging through the snowy morning with chains on their tires, the inside of the massive plant was warm, inviting, and comfortable. The only word possible to describe the Chrysler aerospace division was futuristic. The installation was far beyond impressive.

As Bill walked Darrell alongside the body of a sleek Redstone resting in a movable cradle, a team of technicians was grinding away a section of welding that had not passed an X-ray inspection. As a shower of sparks spread across the polished concrete floor, another team was sweeping up any aluminum fragments that might be left behind. Chrysler was taking no chances; there was even a man with a fire extinguisher, just in case the flying sparks began to cause trouble.

Hinkle spoke over the noise of the grinders and the rumble of an overhead crane. "You'll be working throughout the plant," he offered, "but most of what you will be doing will be in an office and not here on the floor. Guidance is your area of expertise, and with the new and improved Jupiter version of Redstone, guidance at liftoff will be crucial."

With Hinkle leading the way, a nearly completed Redstone was shining sliver as it traveled slowly overhead with canvas slings and chains. "We call the current version of Redstone 'Old reliable,'" Bill said as he pointed to the lifted rocket moving with the crane. "But the new and long-range Jupiter prototype is a whole new animal. Our goal is to transition from the first generation to the next without any problems. That's one of the reasons we need you."

Bill Hinkle smiled as he led the way through a hanger door and into a quiet elevator. When the door closed and the men were moving upward, the Chrysler man offered a sidelong glance. "All of us that have a top security clearance have heard about the German V-2 that you reprogrammed for the Army. That was quite the day in the field of guidance. I hear the Navy is still upset, and a few of the pentagon admirals are taking that little experiment personally."

Darrell shrugged. "There really wasn't anything to it," he said. "Anyone could have done that. It was just an idea. Sperry asked me to solve a problem."

"You are much too modest, Mr. Loan," Bill Hinkle smiled as he stuffed his hands in his pockets. "It's not how you did it; it's how you thought of it. That's what was really impressive. It's the innovative thought and ideas that count."

As the elevator doors opened and two men emerged into a corridor with very white walls and deep red carpeting, Bill Hinkle looked at his watch. "Our timing is perfect," he said. "There is a briefing in about ten minutes and there are some interesting people you'll want to meet. I know this is your first day, but this morning we have a special treat. It's not just a coincidence that you're here today."

After turning a corner, Bill stopped at a door labeled: "Observatory." He then turned and faced the newcomer from Sperry. He was looking for a reaction.

"How would you like to have an up-close-and-personal look at some photographs that were taken of a Russian rocket?" Bill said. "One of the same Russian rockets that lifted Sputnik up into orbit, and the second launch that carried the dog that they killed up in space."

"You're kidding! How is that possible?" Darrell was clearly shocked, and Bill Hinkle was pleased with the results.

"No, I'm not kidding, and *here* anything is possible," Hinkle said as he opened the door. "Welcome to the fine family of Chrysler Missiles."

Chapter 22

*"Science has not yet mastered prophecy. We predict too much for
the next year and yet far too little for the next 10."*

— Neil Armstrong

When Darrell and Bill Hinkle entered the conference
room, they were well above the hanger floor and could see
through a spanning bank of glass windows all of the Redstone rockets that were under construction. All of the new
and shining missiles, however, appeared to be the same, and
Darrell wondered when the new Jupiter project would get
underway.

There were eight men in the glass-paneled observatory
overlooking the factory floor. Two wore suits like Darrell
and Bill and four were wearing the white Chrysler coveralls
that were so apparent in the activity below.

After Bill introduced Darrell to the men from Chrysler, the other two men in suits asked everyone to sit down.
At a table lined with coffee cups and ashtrays, the men in
suits went from man to man and asked to see their security
badges. When it was Bill's turn, he produced his wallet and
removed his Top Secret security clearance. When Darrell offered his Crypto clearance, both the men in suits examined
the credential. After a moment, the bigger of the two men
nodded to his colleague and the smaller of the two nondescript government officials opened a briefcase.

Without any personal introduction, the smaller, thinner
man with a balding head began, "What you are about to
see is classified as Top Secret. The images you are about to

inspect were smuggled out of the Soviet Union well behind the Iron Curtain. Our intelligence resources in Washington have been over each of these photographs carefully, and we now know, for a fact, that the subject matter of these photos is genuine." The man paused and looked around the room making eye contact with each of the Chrysler technicians. Bill Hinkle was given a glance and an affirmative nod and then Darrell the same.

"No mention of these photographs," the man continued, "is to ever leave this room. Everyone here has a need to know what we're up against, and that is the only reason we have traveled here today to present this intelligence. For better viewing, the original photos have been processed into the slides that you are about to see. The details are always better with projected slides."

A sparkling silver screen was pulled down over a blackboard and a slide projector was ready in the center of the conference room. Without any comment, both of the men in suits crossed to the bank of windows overlooking the busy hanger floor and began to pull down blinds. When all of the window coverings were in place, the smaller man turned on the projector and the larger man who was obviously in charge went to the wall switch and turned off the lights. With the glow from the projector, the only light in the room, the smaller man began, "Please stop me at anytime," he said, "if anyone has any pertinent questions."

Above the sound of the fan on the projector, the first slide clicked into place. The projected image was a grainy, black-and-white photo that showed a complex of rocket launching pads. It was obvious the elevated concrete structures were for launching missiles because of the service gantry towers beside each individual launching platform. What was interesting was that the gantries and the raised concrete pads were so close together. In the background, a desolate and dessert-like terrain stretched out to a bleak horizon.

"The first slide," the thin man began, "is of the Communist rocket-launching base in the Kazakhstan region of Soviet Russia. This is where the Sputnik artificial satellite was sent into space along with the diving-bell-type contraption that carried the Russian dog—the dog that the Communists killed while she was circling the earth."

The conference room was completely silent with the exception of the fan on the projector. Clearly, everyone was fascinated.

With another slide clicking into place the thin man continued, "This next image is what we now know to be the transport vehicle that carries the Russian rockets from their assembly site to the launching facility."

The next grainy photo depicted a massive locomotive-type machine positioned on oversized railroad tracks. On the top of the machine, there was a flat deck and carrying surface, and a tall crane-like stabilizing gantry with hanging chains. There were several men dressed in gray coveralls standing on multiple operating platforms, and it was obvious that the machine was in motion.

"Good God! Look at the size of that thing!" This was from one of the Chrysler technicians. "If the transporter is that big, how big are their rocket boosters?"

Without further comment, another slide clicked into place and showed the same machine loaded with a large, bulky, and triangular-shaped rocket. All the images were coarse, and in black and white, but there could be no denying the magnitude of what everyone was watching. The Sputnik space vehicle was huge. The Russian equipment was far bigger than the "Old Reliable" out on the hanger floor, and it was painfully obvious that the Communist booster had more than one engine. The Russian rocket had several engine cones.

"That's incredible," announced another worried voice in the dark, and suddenly matches and lighters flared as the men from Chrysler missiles began to smoke. Darrell lit up and so did Bill Hinkle.

With a haze of cigarette smoke rising into the beam of projected light, the next slide clicked into view. Again, the strangers in the suits were silent as the latest image filled the screen. This time there was a bank of electronic equipment that resembled what would be found inside a complex radio station or radar installation. Standing in front of a multitude of dials and gauges was a thickset man with wire-rim eye-glasses and bushy eyebrows. The severe-looking man was wearing a white lab coat and the men surrounding him were wearing darker versions of the same technician's uniform.

The thin man by the projector said: "We believe the individual in the center to be Sergei Korolev. He is apparently the Chief Designer of all the Communist rocketry and the aeronautical scientist who reports directly to Nikita Khrushchev at the Kremlin in Moscow.

"This Korolev character," the thin man spoke the Russian's name as if it left a bad taste, "has reportedly stated over and over again that he is quite capable of placing an atomic bomb on one of his sputnik boosters and delivering it to anywhere within the boundaries of the United States. His plan is to launch from Russia over the North Pole and destroy New York, Chicago, and Detroit. We now believe that is why there are multiple rocket launching pads . . . so the atomic attack can be simultaneous."

No one spoke, but the fan on the projector droned over the worried silence. The following slide was by far the most dramatic and showed a side view of a Russian rocket ready on a launching pad with the service gantry attached and liquid oxygen vapor streaming down the sides. Even from the distance the smuggled photograph was taken, it was once again clear that the multiengine Russian launch vehicles were gargantuan compared to the current Redstone under production.

The next black-and-white image was the showstopper. When the slide clicked into place, the projector revealed the big triangular-shaped rocket during a launch with all the engines firing. The overall effect was an inconceivable amount of white steam, dark billowing smoke, and bright rushing exhaust flames. As it rose to a height even with the service gantry tower, the final image was made even more dramatic as it was obviously a successful nighttime launch, the bleak and winter-like desert in the background, illuminated by the ignited light at the base of the booster.

With the rising Russian rocket dominating the silver screen, the observatory exploded into a chorus of multiple conversations. One of the men in the dark was impressed with the obvious.

"More than one engine!" he offered loudly. "Of course, that's the answer. The answer to more thrust, boost, and most importantly range. No wonder they can lift and carry heavier payloads."

Another voice even sounded angry, "But how can they control the throttle and flow of the fuel with that many engines—and what the hell are they using for fuel? Can you tell us that?"

When the lights came on and revealed the bigger man standing by the light switch, everyone stood up. "How did you get these slides?" one of the technicians demanded.

The big man shook his head sadly and the smaller man answered immediately, "Gentlemen, I did say pertinent questions. Of course, you all must realize that these photos were gathered by secret intelligence operatives. Operative agents that you don't need to know anything about. Next question?"

"How long ago were these photographs taken?"

"Less than six months," was the curt answer. "Next."

"Do we know how many of these Russian boosters exist?"

"Intelligence reports that there are at least four in reserve. It is estimated that the Communists can produce one completed launch vehicle every four months."

"That's impossible!" the leading Chrysler engineer slammed his hand on the table. "How can you know that? We couldn't possibly produce a booster of that complexity and magnitude every four months! We have nothing even close to that thing."

"Once again gentlemen," the thin man said, "pertinent questions only."

With Chrysler men in white still amazed, appalled, and arguing with the men from Washington and the Department of Defense, Bill Hinkle nudged Darrell and both men headed for the door. With a departing nod toward the bigger government man who was always watchful and silent, the only other men in suits exited the observatory and began walking down the hallway. Before they reached the elevator, a man that Hinkle knew appeared wearing the white Chrysler coveralls.

"I was told to meet you here, Mr. Hinkle—right here," the man said nervously. "You are to call Mr. Keller at once, from the phone in your office to Warren headquarters—as soon as you are finished in the observatory."

"Thanks Fred," Bill said easily, dismissing the messenger. "I'll call as soon as we can get down to the floor."

When the man in the coveralls turned and started down the hall, Hinkle said, "I told you we wanted you right away. Everything seems to be moving very quickly. These people from Washington called last week and said they had something we needed to see. I had them wait until I knew you could be here." Bill Hinkle looked concerned. "Darrell, what did you think about the slides?"

"After watching that, it's pretty hard to think at all," Darrell shook his head. "How could we be so far behind?" he asked pointedly. "As for thinking," he added, "I'll be too busy thinking to sleep—New York, Chicago, and Detroit— all at once? I had no idea that it was this bad—if the public knew about this . . ." he left the rest unsaid.

Chapter 23

"We dare not tempt them with weakness. For only when our arms are sufficient beyond doubt can we be certain beyond doubt that they will never be employed."

— John F. Kennedy

Bill Hinkle clapped a hand on Darrell's shoulder as they walked toward the elevator. "Don't worry too much," Bill offered. "Those guys from Washington are always scary and always intimidating. I think they practice it."

Darrell glanced back to the closed door of the observatory. Clearly, he was concerned with the slideshow from Washington, the surprised reaction of the Chrysler engineers, and with the cryptic message from the man in the hallway. "Those two don't need to practice much more," he said. "They're scary enough—especially, with what *they* have to sell."

Bill laughed. "We can be scary too," he said and his confidence was showing. "You might have noticed that all of the Redstone rockets here are missing one vital element—an element that we talked about at dinner in New York."

"Yeah, I noticed," Darrell said, but his voice was still skeptical. "The important part that *works*: the engine and the range of that engine? Is that the scary part? The range and how high the Redstone can go?" Darrell tilted his head and looked down the hall. "Those technicians in the observatory seemed concerned that one engine might not be enough."

After an uncomfortable moment of silence, the elevator came and Bill Hinkle offered, "Like I said before, at

Chrysler, we do everything in teams. There are no engines here because most of our engine works are in Alabama. That's at the Redstone Arsenal at Huntsville. That's our secret. Wernher von Braun is there with the best of his Germans from the V-2 project."

Hinkle jerked a thumb to indicate the observatory. "I don't care what those defense guys say about the Russians," he said. "We may be behind, but von Braun and his boys are the best. They are about to mount the latest Redstone engine into a brand new and bigger Jupiter prototype booster. He even has some new German technicians that will be arriving any day."

After Hinkle delivered the latest information, the two men rode down in silence until the elevator doors opened on the factory floor. Clearly, Hinkle was preoccupied from the unusual message to call Chrysler headquarters, and Darrell with the thought of newly arrived German technicians—the technicians that could have arrived from anywhere—even from behind the Iron Curtain.

"How many of the original Redstone models have been tested?" Darrell asked over the clamor of grinders, pneumatic tools, and rumbling forklifts. There were easily over two hundred men and women animated in the white Chrysler coveralls.

Bill Hinkle's answer was delayed as he stepped away from a sudden shower of sparks and a noisy grinder. The man operating the machine was oblivious of anything but his work.

With the mid-morning production running up to full speed, Bill spoke above the noise level as the two men began walking toward a section of glass-lined offices located on the factory floor.

"To date, thirty-seven successful launches, and five of those were live atomic bomb tests delivered by our very own Chrysler-built Redstone. Von Braun started designing and fabricating the Redstone missile when the Army brought the Germans from Fort Bliss, Texas, to Alabama in 1950. Before that, and after being sent over from Germany, von Braun and his key technicians were in White Sands, New Mexico, refurbishing and launching the V-2's that had been shipped over after the war." Bill Hinkle winked and tried to lighten the mood. "Sounds almost like what you were up to for the Army against the Navy."

Darrell smiled briefly for the first time since leaving the secret slideshow, but there was a shadow over the smile. "Why did the Army pick Alabama?"

"Funny you should ask," Hinkle was leading the way across and through the production lines. "When the Germans heard that they were going to Alabama, they thought they were going to hillbilly land. They believed they were going to work where no one wore shoes and everyone drank moonshine whisky morning, noon, and night. What really happened was that Huntsville was a munitions factory during the war but was scheduled to close in '49. That was until the Army wanted a place for missile development and rocket engine testing. That's how the Germans ended up there."

Darrell looked up as another shining Redstone began moving overhead. "So von Braun and his Germans have been working in Alabama since 1950?"

"That's right," Bill nodded. While talking, the only two men not in uniform crossed diagonally through the plant and into the quiet of a comfortable office with plate glass windows overlooking the Chrysler technicians in white coveralls. Everything at the plant seemed to be red and white. Bright white walls and ceilings, and in the offices and conference rooms: lush red carpet.

Bill picked up a telephone and said, "Excuse me for a second." After that, he stood at the corner of his desk and dialed in the numbers. Darrell could not help but listen to the one-way conversation.

"Yes sir," Bill said almost at once into the phone and Darrell noticed his voice sounded guarded. "We just finished the briefing."

After a lengthy pause, and a concentrated moment with the receiver, Hinkle continued, "The extended range prototype should arrive in Alabama today, but the snow is bad in Kentucky and Tennessee. The transport boys know to be careful on those icy hills."

Suddenly, Hinkle appeared uncomfortable as he looked over to Darrell and said into the phone, "But Mr. Keller, they just arrived yesterday. They've only been in the house one night."

Bill shook his head. "Yes sir," he said carefully, and his voice sounded even more guarded, "I know this is important and I know about the new schedule. No sir, I did not know you had spoken to the general. Don't worry, we won't let you down. I realize that, of course."

Finally, the Chrysler man shook his head, looked out through the windows onto the factory floor, and then nodded with resolve. "I don't know about the airport," he said. "Yes sir, I'll find out."

After Bill Hinkle replaced the receiver, he said, "That was the boss, Mr. K.T. Keller. Apparently, he has had a few phone calls and even one from Washington. Another call came from General Medaris in Huntsville. Evidently, the general has been talking to Dr. von Braun . . ." Hinkle let the words hang and Darrell looked concerned.

"It seems," the Chrysler man explained, "that the general and Dr. von Braun have also heard about your expertise in guidance and they want you in Alabama right away—as soon as possible. Apparently there are quite a few engineers headed down there right now. There's going to be some kind of briefing and then an engine test."

Darrell shook his head. "Audrey is not going to like this," he said. "Not one little bit."

"I know this is a rush," Hinkle said, "and unexpected. But I have never spoken personally to Mr. Keller, or to the general, or to Dr. von Braun—and Washington?" Bill Hinkle paused to shake his head. "I guess we're not the only ones to have seen the slideshow. It seems that the schedule to get a satellite up and circling the globe just got a lot more urgent and that means all hands on deck—and on the deck now."

"Darrell looked at his watch. His first day at Chrysler was not even into the third hour. "When do I go?" he asked. "And how am I going to get there?"

"You mean, when do *we* go?" Hinkle said and he was suddenly smiling and cheerful again. "I'm going too," he winked. "We are going together—either by airplane, train, or car. Do you want me to call Audrey?" he asked suddenly and his whole demeanor changed. "I feel like this is all my fault."

"No," Darrell said, "I'll call, but I sure wish we could have had a couple more days."

"Don't worry too much," Bill Hinkle offered. "We probably won't be gone long, and besides, I'll send over some Chrysler wives to make Audrey feel right at home. The Chrysler girls are always great and they know just how to stick together. This kind of thing can happen from time to time. "

"It's happened before," Darrell offered carefully, and then he looked thoughtful. "I have an idea," he said. "Do you think you could send the Chrysler wives over before I call?"

Chapter 24

"Efficiency is doing things right . . . effectiveness is doing the right things."

— Peter Drucker

Bill Hinkle certainly was efficient, Darrell decided, and a perfect blend between the two men from Washington that were in charge of the top-secret slide show. Bill was also nondescript. He was of medium build and height and his brown hair was beginning to recede. His brown eyes were always friendly, but there seemed to be something beneath the friendly gaze that spoke of intelligence beyond that of an administrator or an accountant. Bill Hinkle looked as if he could be one of the slideshow men from Washington.

Before leaving the plant at 16-mile, Hinkle made nonstop phone calls with precision and authority: First, to arrange a welcome wagon of Chrysler wives to visit Audrey and to discreetly deliver some spending money. Then, a call to the airport to learn that the snow was still coming down, the winter weather worse, and all the planes grounded. The next call was to the train station in Detroit. In only minutes, the Chrysler man had booked sleepers all the way down to Alabama. A freezing December fog was rolling in with more snow expected, but the southbound trains from Detroit were running right on time.

After the train was booked and Hinkle hung up the phone, he glanced over to his newest travel partner. "I guess the snow is getting pretty bad, "he said. "We had better jump in the car and get down to the depot. We can have lunch while we wait for the train. We leave in three hours."

"We're going to be traveling pretty light." Darrell raised his hands from his sides indicating the two men had no luggage.

"Don't worry," Hinkle said with another wink. "We'll get what we need on the train and a change of clothes in Alabama—toothbrushes, razors, everything. All that we need—Chrysler will cover all our expenses. I've done this kind of thing before."

"I can tell," Darrell said softly, "By the way, how long have you worked for Chrysler?"

"Since the Redstone program began," Bill Hinkle answered evenly. He was already across the office and opening the door that led into the moving wall of sound and the Redstone missiles that were under construction. Clearly, it was time to go.

"Where did you work before Chrysler?" Darrell asked as the two men merged into the arena of productivity and the sights and sounds of America in a race for space.

Hinkle paused for only a second, and then he smiled "Here and there," he spoke over the factory noise. "Here and there."

Chapter 25

"If you want the rainbow, you've got to put up with the rain."

— Dolly Parton

When the doorbell rang, Audrey was surprised. She was certainly not expecting anyone, but when she went to the door and looked outside, there were five young women about her own age standing on the snow-covered steps. They were holding parcels and what appeared to be bags of groceries. All the girls were wearing sensible full-length Midwestern coats and headscarves against the winter wind. Their faces were rosy and cheerful and it was obvious they were friendly and eager to come inside. The snow was still coming down, and Audrey could see that although the driveway and the street were covered in a white December blanket, there were deep tire tracks in the snow and two new Chryslers waiting in the driveway.

When Audrey opened the door, all she heard was a chorus of, "Welcome to Chrysler!" and she was suddenly very pleased that she always dressed well even at home, even if she wasn't expecting anyone, but that was how she was raised, and that's how good girls from the Midwest stayed out of trouble.

"Please come in!" Audrey said and then she laughed. She could not help from smiling. She now knew for certain that she was back in the Midwest. New York women would never be so thoughtful and Audrey was tired of not having girlfriends.

In moments, the girls were inside with a burst of activity and a newly rekindled fire was crackling in the grate. All the women were talking at once and the parcels of everything from chocolate to coffee to apple pie to fresh bread from the baker were piled high on the kitchen table.

"We are so glad to meet another gal and welcome her into our club," announced a pretty redhead with freckles and beautiful green eyes. After holding Audrey at arm's length, she gave her a quick hug and said, "I'm Sally Booth, and if you ever need anything, you call *me*!"

"We're part of the Chrysler Wives Club," a petit blonde happily explained. "Ford and General Motors also have wives clubs, but ours is the best. Our husbands are the best, because they're building the Redstone rockets at 16-mile!"

"That's right," another redhead agreed as she pulled off her headscarf. "Our husbands will be working with your Darrell down at the plant, and we'll be fast friends in no time. We host parties, and do all the fun social events. We even do a cookbook."

"Look at your dress! It's beautiful." the oldest woman broke in. She was an attractive brunette in her late thirties. "My name is Mildred," she said. "Audrey, you are so pretty; I just can't wait to get to know you better."

Before Audrey could reply, the telephone rang. After only a few moments, and very few words, the receiver was replaced.

The Chrysler Wives Club had been briefed. They knew what to expect. All of the wives had seen the same look that Audrey now had on each other's faces.

"Don't worry honey," Sally, the pretty redhead rushed up, "our men are in a race with the Russian Communists. We're all in a race for outer space. It's only natural that the men have to travel and be away. He'll be back soon. You'll see."

"That's right," Audrey admitted. She was determined to be strong and forced a smile. "Men have to travel and be away," she said. "It's happened before, and it will probably happen again. But I don't have to like it—not one little bit."

Chapter 26

"Sometimes questions are more important than answers."

— *Nancy Willard*

At the train station in Detroit, Bill Hinkle paid for the best tickets available, and after a lunch of cheeseburgers and coffee, the two men from Chrysler missiles boarded the southbound overnight train. There were two other men waiting in line for tickets that looked a lot like the slideshow men from Washington, and Darrell thought: *I'm never going to be able to sleep on this train.*

When the whistle blew and the locomotive pulled out of the bustling station, Bill and Darrell were having coffee and cigarettes in the observation car. The snow was still coming down, but now, as promised, an icy December fog was rolling in and turning the touring windows of the sightseeing compartment into big gray lights.

"How did it go with Audrey?" Bill asked carefully.

"About as well as could be expected," Darrell's answer was cautious and it was obvious he was frustrated. "I think it was a very good idea," he said, "to call the Chrysler Wives Club; otherwise, Audrey might have launched her own rocket."

Bill lit a fresh cigarette and looked across the train to the two men who had been waiting in line at the ticket counter. Both were smoking, drinking coffee, and reading newspapers. They appeared uninterested in anything but the newspapers. When the train began moving with a start and a jolt, the depot and the train station disappeared in a swirl of fog.

"How does it feel," Bill asked, "to be traveling down to Alabama to meet the best rocketry engineers on the planet, to know that they asked for you come and help with the race for space?"

"It feels good to be wanted," Darrell said, "or to be needed, but I'm curious, tell me about von Braun. Is he really that good—as good as they say?"

Bill Hinkle eased his shoulders, adjusted his chair, and leaned back against the observation windows. Darrell noticed that although the move was minor, the man from Chrysler could now observe discreetly, out of earshot, the two men in the nondescript suits.

"Von Braun is a genius," Bill began simply, but softly. "And there are some other Huntsville Germans that are just as gifted—one in particular. Some of these guys don't even have university degrees but they are naturally talented engineers—very talented."

"Tell me about von Braun," Darrell asked quietly. "Can you give me an example of what makes him such a genius?"

"Sure," Bill smiled. "I can tell you a story that goes all the way back to the Nazis. During the war, when the V-2 rocket was coming off the drawing board and going into prototype testing, the Germans had a problem. They were using kerosene and liquid oxygen as a primary fuel and oxidizer and pushing the fuel together as fast as possible with a peroxide and calcium fuel pump. The problem began early on in the flight, when the heat inside the engine cone would melt through the metal and send the V-2 flying off course to crash. It didn't happen every time, but it was a problem that needed fixing."

Hinkle paused with his story to watch a conductor moving through the observation car. The man was dressed in a dark blue uniform, with hat, and ticket puncher, and he had stopped to speak with the two men in suits that were reading the newspapers.

After a moment, one of the men opened a wallet and showed something to the conductor. It wasn't clear if it was tickets, money, or an identification credential, but whatever happened instantly satisfied the train official and once again he was moving down the aisle in the center of the train. He did not stop and ask Darrell Loan or Bill Hinkle for anything.

When the conductor was past, Hinkle continued with his story, "All of this happened when the Germans were testing their prototypes at Peenemunde. That was the secret Nazi base in northern Germany next to the Baltic. As the story goes, the rocket engine burn-through problem kept von Braun and his buddies up at night. They had a beautiful rocket that impressed Hitler and all the other big-wig Nazis, but all the Germans knew the problem had to be solved."

"That sounds like a pretty serious flaw," Darrell said, and then he sat back and tried to relax. It was obvious that Hinkle was watching the men in suits from the corner of his eye. It was also obvious that the men in suits, from time to time, were observing the two travelers from Chrysler.

Suddenly, Hinkle seemed preoccupied. His train of thought lost out the window and into the December fog. Darrell probed, "How did von Braun fix the problem?"

Once again, Bill Hinkle smiled and focused, his brown eyes twinkling. "Pure genius," he said. "After a few experiments with different metals, and the results being more burned-out engine cones and out-of-control rockets, von Braun knew he needed an insulator. He devised a plan to use alcohol under pressure flowing down the inside of the engine cone. The part that burned through he insulated with another but very separate flame. The kerosene and liquid oxygen ignition were too hot for any metal so he insulated the hottest flames—the flame that melted the engine cone with another flame. The alcohol flame burning on the inside of the cone was the flame with the lowest temperature and *that* flame protected the metal that had to contain the heat of the combustion. Surrounding a flame that was too hot with another flame that burned cooler . . . that's what I call genius."

In an instant, Darrell forgot about the mysterious men in suits, Audrey launching her own rocket fueled with anger and frustration, and the urgent schedule that Chrysler and the U.S. government were about to unfold. Darrell was beyond impressed. "That's why the Redstone we launch today uses alcohol for fuel instead of kerosene?" He smiled in understanding and said, "That *is* true genius."

Bill nodded, "Yes, and that's just the beginning and only one example. This guy comes up with ideas like that on a daily basis."

"So tell me," Darrell leaned forward and asked quietly, "Were von Braun and these other Germans really Nazis?"

Hinkle grinned. "You bet," he said. "You don't get that kind of funding without being on the political bandwagon. They were all Nazis—all those German rocket scientists—but they had to be. We really can't imagine what it was like, living under Hitler, but not being a Nazi for the German elite during World War II was like driving a car without a driver's license."

"But what about those war crimes trials at Nuremberg?" Darrell was still leaning forward, his voice out of earshot from the men in suits. "What about that?"

Bill shrugged. "It's all politics," he said. "Besides, our Germans were scientists and technicians and not responsible for mass murder."

"Tell that to the families in England and Holland," Darrell said, "When those Nazi rockets delivered a ton of high explosive without warning at the speed of sound. A lot of people were killed in those attacks and their relatives will never forget."

Suddenly, the man from Chrysler stiffened. The two men in suits were standing. They were approaching and moving down the aisle. When they were past without a greeting, nod, or smile, Hinkle offered: "I have a feeling those two are FBI and it's no coincidence that J. Edgar Hoover's men are on our train."

Darrell shrugged. It was going to be a long night.

Chapter 27

"Freedom lies in being bold."

— Robert Frost

As the train rumbled south through the fog and the snow and the early nightfall, Darrell Loan and Bill Hinkle had dinner in the dining car. Because the two strangers in suits were at the next table, the conversation at dinner was all about basketball. The men in suits didn't seem to be having any conversation at all; they seemed to be listening. After dinner and drinks, Darrell climbed up into his sleeper. To his surprise, he slept like a baby.

Early the next morning, the locomotive pulled into Huntsville and found the train station packed. It was clear that something important was underway because it was painfully obvious that Huntsville was a small southern town and could not warrant all the bustling activity. For many on the train, Huntsville was the final destination. When the doors to the passenger cars opened, the depot was milling with new arrivals and military personnel. Across from the landing platform, army staff cars were waiting and lined up outside the station.

After Bill Hinkle spoke to one of the drivers with an empty car, the soldier smiled, nodded, and opened a door. Before Bill and Darrell settled into the back, the two men in suits from the train rushed forward and signaled the driver to wait. The man in the lead was puffing big clouds of winter vapor into the crisp morning air, and after a moment, the two strangers from the train crowded into the olive green sedan with the white star. The inside of the car was over-heated and close.

"You guys with Chrysler, right?" the stranger crowding Bill Hinkle in the backseat asked. The man in the front seat only looked forward as the car pulled away from the station.

Alabama was cold and bleak in winter, the trees bare, but there was no snow. The morning sun was wintry and weak with dark clouds climbing over the railroad tracks and gathering on the horizon.

"Yeah, we're with Chrysler," Hinkle offered at once. Then he asked abruptly, "You guys FBI?"

The man in the back seat laughed, "No," he said, "we're with Chrysler too." Everyone laughed, including the driver, and after a few minutes, introductions were made and the mysterious men in suits were revealed as fellow engineers from Detroit.

After the army sedan drove through the deep southern town of red-brick buildings and southern-plantation-style architecture, the Redstone Arsenal came into view. From the outside, the base was typical military and everyone was required to show security credentials at the guarded check-point gate. Bill Hinkle and the two older men that were the strangers in suits had "Top Secret" badges. Darrell's alone was "Crypto."

With all the credentials cleared, the army driver drove through a series of well-kept military streets. There were soldiers everywhere puffing through the cold winter morning, and behind several older barracks, there was a view of many large and newly built and very large hanger-type buildings rising in the background.

The driver stopped in front of the steps and facade of a new and modern building. There were military police guarding either side of the wide main entrance, and above a glass-door vestibule, there were two gunmetal signs with white bold lettering. The uppermost sign declared:

Headquarters U.S. Army Ordnance Missile Command

Below the first was another equally impressive and clinical looking sign that offered:

Headquarters U.S. Army Ballistic Missile Agency

On either side of the overhead portico and on the shoulder of every soldier visible was an emblem and a mission patch. The shoulder patch insignia was an amber rocket on a background of red and blue with silver lightning bolts on either side of the rocket. The rocket on the crest looked just like the Redstone models Chrysler was building in Detroit.

Inside the administration building, there was a long desk and a sign that offered: "New Arrivals." After waiting with about thirty other men all dressed in civilian suits, an army captain in a pressed and starched uniform appeared at the reception and asked for quiet.

"Gentlemen," the captain spoke with a strong southern accent. "There will be a briefing in the conference room at ten hundred hours. Everyone here should sign the roster at the desk, and everyone that signs in is required to attend the meeting. Ah, thank you."

Darrell looked at his watch. "That's ten o'clock, military time," he said. "Just under an hour."

Bill Hinkle nodded. "I guess we're in the Army now, and in the south," he offered. "Like it or not."

Before Darrell could comment, Hinkle chanced a sidelong glance at the two strangers from the train. "By the way," the Chrysler man covered his mouth as he spoke, "there's no way those two work at Chrysler."

Chapter 28

"If we open a quarrel between past and present, we will find that we have lost the future."

— *Winston Churchill*

The central Redstone conference room resembled a large classroom with a blackboard and a pull-down movie screen behind a heavy wooden desk. Chairs were set out in neat military rows facing the desk and the blackboard. As the civilians began filing into the room, everyone began taking seats. It was clear from the start; everyone wanted a chair in the front row. With the front seats already taken, Darrell and Bill sat with the strangers from the train in the second row.

Just as Darrell was wondering if there was going to be another scary slideshow with a projector, the southern army captain appeared at the door and crossed over to the desk. There were two other men in dark blue suits with very white shirts and dark ties standing near the door. Both had appeared with the captain and now stood quietly with their hands in their pockets.

"Good morning and thank you all for coming," the captain announced as he faced the seated group, his southern accent even more pronounced. "My name is Freddy Shultz and I am the chief liaison officer for the gentlemen that are about to be introduced."

The captain paused for effect. The ironed creases on his uniform were perfect. At this time," he offered, "I would like to introduce the chief designer of the Army Ballistic

Missile Agency and the chief advisor to the Army Ordnance Missile Command: Dr. Wernher von Braun."

As the immaculately dressed von Braun advanced from his waiting position by the door to stand behind the desk, a pencil falling to the floor would have sounded like an avalanche. All of the newcomers remained absolutely quiet, and there was at once, Darrell decided, a feeling of uneasiness and mistrust throughout many of the older men in the room.

Von Braun had been a Nazi, an officer in the dreaded SS, and he had waged war on the United States only twelve years earlier. It was clear that many of the Americans in the room had fought against the Germans in World War II, and many of the men seated and facing the German had lost friends, relatives, or comrades to the Nazi war machine.

"Good Morning," von Braun said thickly as he smiled. "I think, I believe," he hesitated, "that soon . . . we will all . . . be very happy in our work together." The words were heavily accented with a very strong German inflection, and it was clear that von Braun was thinking in German and then translating his thoughts into English.

No one said a word and the silence was painful. The southern army captain that had introduced himself as Freddy Shultz was standing near the door with von Braun's obvious cohort. Both the soldier and the civilian were curiously overlooking the new arrivals and evaluating the stoic situation. The seconds passed slowly.

Abruptly, von Braun nodded, stuffed his hands back into his pockets, and began to pace behind the desk.

"Gentlemen," he began again, "you have been chosen because of your talents and advanced aptitude in engineering. Your—" he hesitated, obviously thinking for the word, "—your skills—have already been proven, or you would not be here today . . . listening to my dreadful English."

Von Braun looked up as a few muffled laughs broke the tension. The German instantly smiled, and his smile was at once winning and charismatic. Suddenly he was boyishly good-looking and didn't look old enough to have been in the war.

"Yes," he continued, "I am honored that we will, all, be working together, but I must explain that we are behind schedule and therefore under a great pressure to produce what America needs badly." He paused to look around the room and his gaze was piercing. "Ja, Ja, a great pressure to prove that America and the free world have the technical expertise to be superior to that of the Communists in Russia."

Von Braun paused. Clearly, he was searching the faces of the men seated before him. He was looking for a positive reaction. He was hopeful for trust.

"If the Communists continue with their conquest of space, and we are unable to respond with our own efforts, I think the world will soon believe that the Russians, and perhaps the Russian way of life, is more advanced and perhaps superior."

Wistfully, Von Braun shook his head. His dark navy suit was pressed and his tie was perfect. Suddenly, his hands came up dramatically and his suit jacket was showing the whites of his cuffs and his gold cufflinks. His arms were raised as if to embrace the room.

"Of course," he said thickly, as he smiled his winning smile, "there is no chance that the engineers in this room today will allow the Soviets to win." He nodded with confidence and quickly continued, "Our newest long-range Redstone will lift our satellite successfully, and the Russians and the world will know," he gestured again to include everyone that was watching, "that the American effort into space will not be surpassed by the Communists."

"But I must bring forward a warning to everyone here," the German said sternly as his hands came down. "We all have a great deal of work ahead, and we must press ourselves very strongly in the days and weeks to come to make our success complete."

"I will now introduce," von Braun's arm was en-gesture to bring forward the man standing next to the army captain at the door, "Herr Doctor, Kurt Debus."

As the other man approached the desk, it was clear that the two Germans wore their immaculate navy blue suits as the newest uniform of the day.

When Debus stood beside von Braun, he said simply and offered to the room, *"Guten Tag"* — good day in German.

Von Braun smiled. "Dr. Debus," he said, "only speaks the language of his fatherland." Quickly, von Braun translated what he had said for Debus and momentarily the room was spellbound as the Americans listened to the rapid-fire German exchange.

"The *Herr* Doctor has asked me to explain," von Braun said in his heavily accented English, "that he is looking forward to our interesting work together. I feel I must add to the *Herr* Doctor's statement by saying, I believe that Dr. Debus is perhaps the greatest electrical and mechanical technician I have ever had the pleasure to work with."

Debus smiled and nodded. He was shorter, thinner, and older looking than von Braun, but the intensity and intelligence behind his eyes was unmistakable. His dark hairline was receding and there was a very distinct and savage scar on the left side of his chin.

"Zo," von Braun said with his accent and reclaimed all attention. "We will have a static test firing of the Redstone engine we will be using in the new Jupiter booster at 1500 hours today."

"I must insist," von Braun again raised his arms to the room, "that everyone here will be in attendance. I can also make a promise," again the charismatic smile, "that no one will be disappointed . . . or unimpressed."

Chapter 29

"Even if you're on the right track, you'll get run over if you just sit there."

— Will Rogers

Just before 1500 hours — 3:00 p.m. Huntsville time, all the new arrivals and the engineers that had traveled to the Redstone Arsenal during the previous days were assembled near a large open hanger.

The newest Jupiter engine prototype was mounted to a concrete test platform with three strain-gauge clamps attached to the base of the engine exhaust cone. Two army-green tanker trucks were parked at a discreet distance near the open hanger with thick hoses running over the ground to connect with the intricate tubing at the head of the rocket engine. The fuel truck closest to the hanger was labeled "LOX" for liquid oxygen, the second tanker "Methyl" for methyl alcohol. Several bundles of wire cables led from the engine head and snaked along the tarmac to disappear inside the open doors of the hanger. Two army brigade fire trucks with two crews of army firefighters were busy uncoiling and connecting water hoses to nearby fire hydrants.

Most of the civilians were wearing overcoats against the chill of a brisk northwesterly wind, and the sky was low, gray, and completely overcast. As the new arrivals milled around the horizontal rocket engine and examined closely the complex tubing and details of the powerhouse for the newest Redstone booster, everyone was aware of the over-sized "No Smoking" signs posted by the fuel trucks and the engine test stand.

"Does this look familiar—anything like the V-2?" Bill Hinkle asked as he stamped his feet against the cold. Both Darrell and Bill were wearing borrowed army-issue great-coats over their suits they wore from Detroit.

"No, this looks nothing like a V-2," Darrell offered, "and I've never seen an engine test, but I'll bet it's going to be dramatic. Look at that burned-up concrete."

Before Hinkle could reply, the army firefighters opened valves on the water hydrants and the empty canvas hoses charged into life. The initiating Huntsville fanfare began with a great stream of hissing water. The first powerful stream arched out over the burned and cracked concrete where the conical exhaust cone waited. As the second fire-fighting hose jumped to life, it was clear that the water pressure was impressive as it took two firefighters to hold each of the fire hoses steady.

After several minutes of the arching wind-blown water, and after the burned crater that had once been a concrete tarmac was filled with the overflowing deluge, an army staff car with a white star approached, and von Braun, Debus, and the southern army captain emerged with coattails flapping in the winter wind. The time was exactly three o'clock.

After a quick glance at his watch, Bill Hinkle nudged Darrell and said, "Right on time."

With his arms en-gesture von Braun was motioning everyone to come closer. He was animated with his boyish smile and standing beside his rocket engine. Before he spoke, both of the gushing fire hoses were wrenched into silence and all the firefighters stood quietly with practiced reverence. It was suddenly very clear: the men in the army fire brigade were waiting for the German to speak.

"Thank you for coming," von Braun announced as all the new arrivals gathered closer and the firefighters strained to hear. "With the extended fuel tanks and stages needed to achieve orbital flight, our standard Redstone engine has been modified. We have calculated that the fueled booster and engine will have a combined launching weight of 66,000 pounds. Of course, this figure does not take into consideration the added weight of the satellite spacecraft. Tolerances at liftoff will be crucial. I do feel confident, however, that our newest endeavor will correctly lift a substantial payload into orbital space." Von Braun paused smiling, and then spoke suddenly to Debus in German.

After Debus nodded quickly in understanding, von Braun continued, "And now I will ask that we all take our safety positions please."

Freddy Shultz, the southern liaison captain that had emerged from the staff car with Debus and von Braun took the lead. He quickly crossed over the fifty yards of wet tarmac and water puddles and led everyone to stand behind a makeshift wall of sandbags that reminded Darrell of his launching experience with the V-2.

Before everyone was behind the protective barrier, Bill and Darrell were surprised when they heard the U.S. army captain, Freddy Shultz, the liaison officer with the strong southern accent, speaking fluent German with Debus and von Braun. After the German exchange was finished, von Braun faced the newcomers.

"Prepare yourselves," he said proudly, "for precisely three minutes . . . I think . . . winter will be over." He then nodded to the captain and Freddy Shultz raised his arm to signal a soldier waiting beside the firefighters in their safety gear.

At once, a red star shell was launched from a signal pistol and all eyes went aloft to follow the brightly burning flare. After the warning flare arched and was finished, the winter day vanished with a tremendous burst of heat, flame, and a thunderous roar. The Jupiter Redstone engine ignited as most were watching the safety signal, and the rocket engine start caught almost everyone unaware.

The incredible heat flash that leapt across the distance to the sandbag wall was so intense and instantaneous that many of the new arrivals were unprepared and ducked behind the fortified refuge. Abruptly following the heat flash, an earth shaking vibration and sound wave began as the new Jupiter engine was consuming hundreds of pounds of fuel and oxidizer with every passing second. As the unbelievable ripping sound of the thundering combustion continued, it was difficult to decide if the all-consuming vibration was more demanding than the unwavering wall of thunder.

Von Braun, Debus, and the southern army captain had remained standing and watching, as did Bill, Darrell, and a few of the other engineers, but most were only rising to observe the static test after they understood there was no inherent danger. The rocket engine had ignited perfectly, and as it continued to throttle up to full power, the overwhelming noise level increased as did the flame exhaust that easily reached out from the engine cone for over a hundred yards. Upon the Jupiter engine's ignition, the firefighters had gone into action, and as the startled engineers watched — some with their hands over their ears — the fire brigades sprayed water over the concrete that was under attack from the liquid oxygen and alcohol-driven exhaust flame.

Even with the intense heat, flame, and chest-pounding vibration, it was obvious the Germans and the army captain were attempting to communicate. The trio was in a huddle, their hands moving as they nodded their heads. After it was clear that the combustion was at full run-up and operating perfectly, von Braun, Debus, and the army captain were looking and smiling toward the assembled engineers that were clearly awestruck.

When the Jupiter engine abruptly shut down, a deafening silence prevailed and was only intruded upon by the gushing fire hoses spraying the steaming and smoldering remains of the blackened tarmac.

At once, von Braun, Debus, and Captain Freddy Shultz were in deep Teutonic conversation and looking at their wristwatches. By their reactions, the three-minute time sequence for the static test had gone perfectly.

"Damn!" Bill Hinkle said over the silence and the unstoppable German exchange. "I never expected anything like that. What power! That was unbelievable."

Darrell's smile was unstoppable, "Bill," he said, "I sure am glad you called. I'm really very pleased to be working for Chrysler. We've got a rocket engine that's going to work."

Suddenly, all the engineers were talking at once. Some were laughing, others clapping each other on the back, but all were smiling with an unbelievable and previously unknown pride and confidence. Assembled behind a sandbag safety wall were some of America's top engineers, and every man present now knew the battle for the conquest of outer space was really underway.

Chapter 30

"Crash programs fail because they are based on the theory that with nine women pregnant you get one baby a month."

— Wernher von Braun

In the days that followed the dramatic static test, the proven Jupiter engine was installed in the new long-range Redstone fuselage that had been shipped down from Chrysler.

Immediately after the engine burn, all of the newcomers had received their assignments. Darrell had been given the blueprints and schematics for the newest Redstone booster, and during the time it had taken the technicians to install the Jupiter engine, the former troubleshooter from Sperry knew almost everything about America's first attempt at achieving an orbital flight. Because of his expertise with compass systems and gyroscopes and his success with the Army and redirecting the V-2 against the Navy's sparrows, Darrell's official assignment in the quest for outer space was guidance.

With a Christmas tree decorated and glowing with bubble lights in a corner of the Redstone recreational hall, Darrell sat alone at a table and looked as his blueprints. There were three sets. He not only had the technical drawings for the new long-range Jupiter booster but copies of the "Old reliable" Redstone and the original German V-2 as well. As the hour approached midnight, he was comparing the newest Redstone with the war-era V-2 rockets that were von Braun's first major success.

The guidance systems were similar, Darrell decided, as much as newer car models had the same basics as those of motor vehicles invented decades earlier. The V-2 was clearly the forerunner, and not much had changed as far as the method for guiding a rocket. The only real and unpredictable problem was controlling the rocket at liftoff. After that, when the booster came up to steering speed, the rudders on the rocket fins would guide and control the flight with the strict control of gyroscopes in the instrumentation package. This, of course, was unless something else went wrong other than guidance.

As much as rocket flight seemed complicated, it actually was not. Just as with the V-2, the guidance of new Jupiter relied on the three basics: pitch, yaw, and roll.

Darrell remembered explaining the principles to Bill Hinkle earlier and was surprised that the concept had previously eluded the Chrysler man. But to be fair, Darrell considered, Bill was not an engineer but an administrator. Even so, it was a pleasure to watch as he suddenly grasped the fundamental and smiled with amusement when he understood the basics that beforehand had seemed so difficult.

"It's really simple," Darrell explained as Bill sat beside him earlier. "There are only three directions that an airplane or a missile can travel — or a boat for that matter." At once Darrell decided that a boat would be the best example, and he quickly sketched out a penciled boat on a piece of drafting paper.

"If you can control the three directions of movement," Darrell tapped the pencil next to the boat. "You can control the direction of anything — even rocket flight."

"The first is pitch. Pitch is dangerous when anything guided wants to hobbyhorse, or in the worse case of pitch, to tumble end over end. Imagine a boat in rough seas when the bow rises and the stern drops over and over again. That's pitch. A boat pitches in rough seas when the nose of the boat goes up and down. We can control the pitch of a rocket with a gyroscope that keeps the pitch angle steady by controlling the steering fins that guide that axis of flight."

"Next is yaw. Yaw is when the same boat without someone steering wants to keep going from left to right over and over again—back and forth. We control yaw in a rocket with another gyroscope connected to another set of steering rudders that hold steady that angle of flight."

"The last and easiest variable to control in rockets is roll. In rocketry, we love roll."

Darrell laughed at the thought. "Roll adds stability," he explained. "In a boat or a ship, roll happens when our same boat is tossed from side to side rolling in the seas. If the roll in a boat weren't corrected, the boat would roll over. In rocket flight, we want the rising missile to continually roll at a speed we can control with another guiding system. Imagine a drill bit boring into something. The rocket is our drill bit, and we want it to bore a hole into the sky." Darrell grinned. "Pitch, yaw, and roll. It's simple."

The look of comprehension on Bill Hinkle's face when the boat and the drill bit lesson took hold was well worth the explanation, but now as Darrell sat alone with the three sets of rocketry designs, he had doubts about the latest technical drawings that were already a reality and waiting to be tested in flight.

Once again, the problem was going to be at liftoff, with a heavier-than-ever booster, carrying an untried load of fuel, and a slow rise to airspeed. Before the Jupiter was climbing fast enough for the steering rudders to control pitch, yaw, and roll, carbon steering vanes were positioned in the engine exhaust to guide the thrust of the rising booster. This was critical, until the speed of the rocket flight was fast enough for the steering rudders to take over.

For the third time since he was alone in the recreation hall, Darrell shifted his blueprints and went back to the original V-2 drawings. He was considering with scrutiny the steering fins on von Braun's wartime creation. There was something wrong, something that didn't make since. After a moment of contemplation, the engineer from Iowa looked up.

The lights in the library-style recreation hall were dim with the colorful Christmas tree glowing nicely. Positioned on the reading table with the blueprints were two desktop lamps illuminating the technical drawings. From the far corner of the wood-paneled room, a radio was softly playing Elvis Presley's latest hit: *Blue Christmas. . . It's going to be a blue . . . blue . . . Christmas . . . without you . . .*

Just as the song reminded Darrell of the snowy Christmas scene in Detroit and a troubled Audrey alone and many miles away, there was a noticeable set of footfalls on the hardwood floor.

"Zo," Dr. Wernher von Braun said softly as he appeared in the lamplight, "it seems . . . I'm not the only one working tonight."

"I didn't know there was anyone else still here," Darrell stood from his chair.

"*Ach*, you can always find me nearby," the German accent explained. "When you are in love with your work, you are not really working. You are just enjoying your pursuits — your sweetheart."

Von Braun offered his hand and Darrell shook it: "I'm Darrell Loan, Dr. von — "

Von Braun smiled in the lamplight. "Of course, I know who you are," he said. "I have studied the records of all our new technicians most carefully. Your exploits precede you."

"How's that?" Darrell was at a loss.

"Come now," Von Braun was smiling his winning smile. "You have been the topic of many conversations. Zo," he continued, "do you imagine that even down here in hillbilly land we have not heard of your success with my old friend the V-2?"

"It was just an idea," Darrell said. "It was something the Army asked for."

"Yes, of course, but it worked and the function was perfect." Even at midnight, von Braun's suit appeared pressed, but now as he slipped out of his suit jacket and looked down at the mechanical drawings on the table, he was ready to go to work. All of the blueprints were aligned and focused on the base of the rockets. He pointed to the tail section of the V-2.

"There is perhaps something you are finding interesting," he said. "Something in which . . . perhaps . . . I can be of help?"

"Yes, yes there is," Darrell nodded. "I have been over each of these drawings again and again, and I can't figure out why the tail section is designed with such short fins and oversized steering rudders. It seems that the flight controls would have worked better if the tail fins and the rudders were longer."

Abruptly, von Braun began to laugh. It was a big belly laugh and full of good humor. "Ach Zo," he said, still laughing, "no one has ever asked such a question—such a very obvious question . . . no one, until now."

"Did I say something wrong?" Darrell was perplexed. "Is there something I missed?"

"You have missed nothing and seen *almost* everything. The reason the tailfins of the V-2 are so short is because the final assembled rocket had to travel through railroad tunnels. The originals, as you have speculated, were indeed longer, but they were too big to travel through the underground tunnels. We were forced to redesign the tail fins for the German railroad authorities. We had to work within certain dimensions. The V-2 with the original long tail section would not have passed through the railroad tunnels in Germany or France."

Suddenly, Darrell smiled in understanding and von Braun clapped him on the shoulder. "All the war efforts were compelled to travel through the tunnels," Von Braun declared, "And on occasion, during the bombings, the railroad tunnels were used as bomb shelters. The tail design was one of necessity."

"I should have thought of that," Darrell confessed.

"Nonsense," the German shook his head. "You could not have foreseen such a situation, but you did think to ask."

"I think there will be a lot of things," Darrell said, "that we will need to ask."

"Yes of course," Von Braun's smile was back, "but for now, I think, our work tonight is finished. We are traveling tomorrow, and we must have our rest. With the morning, we leave with our new Jupiter booster and drive to New Mexico. Tomorrow we start the journey to the White Sands Proving Grounds. We are going to flight test our latest work."

"I didn't know I was going," Darrell confessed.

"Oh yes," Von Braun was buttoning his suit jacket. "I must insist you are present. You might have more questions that we need to answer. I am very pleased that you are here. Now off you go, and good night. Sleep well."

After the newest guidance technician gathered his three sets of drawings, placed them into individual folders, and started for the door, Wernher von Braun paused to look at the glowing Christmas tree. When the young engineer was down the hall and the German was quite alone, he shook his head and offered aloud to the room "Redirecting the guidance of a V-2 to fly horizontally . . . how extraordinary."

Chapter 31

"Success is the ability to go from one failure to another with no loss of enthusiasm."

— Winston Churchill

With an incredible spangle of stars still bright in the clear December morning and a crescent moon setting in the west, the final preparations were underway for a military convoy. The new Jupiter Redstone, assembled with the recently tested engine, was strapped to a flatbed tractor-trailer and covered with lashed-down tarpaulins. Two dusky army trucks and two jeeps with soldiers were stationed before and behind the new Redstone booster, and there were two new Chrysler sedans lined up and waiting behind the military. As Darrell and Bill Hinkle approached, von Braun, Kurt Debus, and Freddy Shultz — the army captain with the southern accent — were in deep conversation. They were speaking German.

Upon recognizing the newcomers in the pre-dawn light, von Braun switched to English and called for the driver of the flatbed. When the solider with the rocket patches stitched on his shoulders climbed down from the truck cab, von Braun motioned him into the assembled group.

"Gentlemen," the German began," I would like to introduce Sergeant William Hammond. Will, as he prefers to be called, has been chosen as the principal driver to transport our newest work."

After Bill Hinkle and Darrell met the sergeant, and everyone shook hands, von Braun continued: "I think you might enjoy traveling with our transport driver. Will has been with us for several years and has many colorful exploits regarding our earlier efforts. Many of our tests have been successful, but some were not. Zo, I think," Von Braun paused and glanced at Darrell, "that the failures might have a special interest to our newest guidance technician."

From the front of the convoy, a military police officer approached and switched off his flashlight. "Dr. von Braun," he said, "we are ready when you are. We'll have sunup in about twenty minutes."

Abruptly, von Braun nodded and said, "Of course." He then turned to follow the mystery that was Freddy Shultz, Kurt Debus, and another soldier as they took their places in the leading Chrysler. Before the overhead dome light winked out, Darrell was amused to see von Braun and Debus already in conversation in the backseat with the liaison captain as the front passenger and the other soldier driving. More civilians in suits piled into the other sedan and with a nod from the military police, Bill Hinkle and Darrell climbed up and into the big truck cab.

As Hammond found the gears, and the big diesel and flatbed began to move out, the sergeant whistled tonelessly. After a few minutes the sunrise was complete, a new day began, and the Huntsville convey was away from the security checkpoint and out on the road to Texas.

"You boys ever been to El Paso?" Hammond asked as he drove.

"No," Darrell and Bill answered together.

"We'll, let me tell you," Hammond said with a sidelong glance, "it won't be like anything you expected."

With the throaty diesel rumbling up to speed and following the winding road, Hammond looked ahead and regarded the escorting military. "We stay in El Paso about sixty miles from The Proving Grounds," he explained over the road noise. "That's in White Sands, New Mexico. The Mexican food and the beer is good, the Texas barbecue great, but the motels we stay in have cockroaches big enough to put saddles on. Texas-sized cockroaches."

Hammond looked mischievously across the cab. "For entertainment, all you have to do is turn out the lights for a few minutes, then turn 'em back on and damn! You've got your very own cockroach rodeo right there in the room."

After Darrell snorted and Bill laughed, the sergeant with the rocket patches continued. "Of course most of the time von Braun and the other krauts don't stay where we do, near the edge of the dunes, probably because they stay somewhere nicer, but maybe because the krauts are careful. There's a lot of peculiar things that happen out in that desert" Hammond paused for effect. "That's right," he nodded, "mighty peculiar."

Bill Hinkle sat in the middle next to Hammond, and Darrell next to the passenger window. Sergeant Will Hammond was a tall thin man with blue eyes and blond hair, and an ever-ready smile. He was instantly likeable, an obvious von Braun and Huntsville favorite, but as he changed gears with his subject matter, a darker side began to emerge even in the bright Alabama morning.

"What kind of peculiar things?" Bill Hinkle asked and lit a cigarette. The clear December morning was still cold, but the cab heater was on, and Darrell and Hammond rolled down their windows for the fresh air.

"Hell and damnation," Hammond lit his own cigarette with the flick of a Zippo lighter. "I wouldn't know where to start," he said. "They say those A-bomb tests have caused all kinds of problems—problems that the brass in Washington doesn't want us to know anything about."

Darrell grinned privately and nudged Bill Hinkle. "Dr., er, von Braun," Darrell said carefully, "mentioned you could tell us about some of the booster failures—something we might be able to learn from."

"There's been plenty of failures, that's for sure." Hammond nodded sagely. "And those too, have been strange and peculiar."

"What would you say has been one of the strangest?" Bill Hinkle asked over the diesel rumble. His voice sounded worried.

"Well," Hammond flicked his unfinished cigarette out the window. "I guess that graveyard down in old Mexico would be one. That was back when we were testing the Wac-Corporal. That was one unstable son-of-a-gun. It was fast though, and mighty dangerous. That was the experimental sounding rocket that used red fuming nitric acid and pure aniline. Both of those fuels were deadly poison, hypergolic, and tricky to handle. That might be why we had that accident in the graveyard."

With Darrell and Bill Hinkle both hooked, landed, and riding quietly, Hammond shifted gears on a downgrade and continued the story: "That was one crazy morning. We were all set with that Wac-Corporal fueled and ready, and when the boys lit that one off, we all got worried real quick. It just didn't act right, right from the start, probably because there was too much smoke. Most of us think it was leaking that red fuming nitric acid. It went up like normal — except it was smoking like hell — but when it was out of sight, two of the boys came running out of the blockhouse and yelled for us to all take cover. They acted like it was coming back — right back to where it started — and coming fast!"

Hammond laughed nervously. "I can tell you we were all pretty scared, and that little number wasn't anything like what we got behind us. This thing," Hammond jerked a thumb backward to indicate the newest Jupiter Redstone on the flatbed. "This one here," he tilted his head, "this will be the biggest thing ever tested at White Sands."

"What about the graveyard?" Darrell leaned forward to ask, "The graveyard in old Mexico?"

"Well, that was just it," Hammond confided. "That Wac-Corporal didn't come back for us directly — we all saw it coming back — but then that fast trail of smoke just turned south and headed off quick. The firing range officer hit the self-destruct over and over, but nothing happened. That little rocket went straight over the border into Mexico and blew up a Mexican cemetery. I hear tell, there was one hell of a crater, and coffins and old bones scattered all over the place. Yep, mighty strange and very peculiar. We all wondered why it went south and headed for the graveyard."

"Did they ever find out what happened — " Darrell asked, "about what happened with the guidance or the self-destruct?"

Hammond shook his head. "The official report, I never got a look at, but I heard it was two crossed wires that made it go off course. But that doesn't explain why the self-destruct didn't work, or why it started leaking that red fuming nitric acid. Everything we launch is supposed to have a pretested self-destruct so if the guidance goes bad we can blow it up before there's any chance it can get away and hurt somebody. We had a lot of those Nazi V-2s go haywire too, and there's a lot of big and burned-out holes on that desert floor."

"Damn," Bill Hinkle said. "What happened with the cemetery? What happened afterward?"

Hammond shrugged. "Everybody says that was one expensive rocket test, because just south of the border there's a lonely little Mexican graveyard that was paid for by Uncle Sam. Paid for again and again, because of that crazy rocket, and because the Mexican government is still on the warpath about that one. Just imagine . . . all those broken and burned coffins . . . with all those skeletons still in the clothes they were buried in . . . scattered everywhere . . ."

"What else peculiar has happened out there?" Darrell interrupted, but after he asked, he wasn't so sure he really wanted to find out.

"The sandstorms can be strange, fast, and deadly," Hammond continued somberly. "One minute the White Sands can be as clear and fair weather as anyone could imagine and the next a big dark cloud will start up over the horizon and blow up into something awful. Sand so thick you can't get a breath and sand so fine it will go through an air filter and choke out a carburetor. People can get lost in storms like that and we've had a lot of those storms lately. All the locals and the old Indians say the storms were never as bad as they are now and some people have disappeared. Most folks believe it's because of our rocket testing and because of all those A-bomb tests."

Suddenly, Hammond leaned forward from his driving position and looked across to his passengers. "You boys ever heard about Roswell—Roswell, New Mexico?" he asked. When the two men from Chrysler shook their heads no, Hammond settled back in his driver's seat and began the story about a New Mexico rancher that had found a crashed flying saucer.

"They say there were little gray men from another world that were watching the A-bomb tests back in '47. Most folks think they were from Mars." Hammond was now into his element. "We've all heard that it's all top secret, but the air force figures those little gray pilots must have gotten too close to the atomic bomb and crashed. There's some kind of magnetic pulse that comes out of an atomic explosion. Something you can't see . . . something that made them crash."

After another moment of road noise Hammond repeated his warning, "A lot of strange things out in that desert, mighty strange and unearthly. Who knows," the sergeant then shrugged and downshifted. "Maybe more of those little gray pilots will come back to have a look at what we are hauling behind us — especially if we can get this old Jupiter up and out over the dunes."

Once again, Darrell grinned and nudged Bill Hinkle, but Bill Hinkle didn't move and he definitely wasn't smiling.

Chapter 32

"Be who you are . . . and say what you feel . . . because those who mind don't matter . . . and those who matter don't mind."

– Dr. Seuss

With the long drive over, and Mississippi, Arkansas, and most of Texas behind them, the Huntsville convoy arrived with the early morning sun at El Paso, Texas. Two Chrysler sedans were already waiting at the now infamous Sombrero Motel as Bill Hinkle, Will Hammond, and Darrel Loan crossed over to the adjacent Mexican-style coffee shop stretching their legs. Everyone was more than ready for breakfast.

El Paso was dusty, dry, and warm even in December, and as the Huntsville soldiers took turns watching over the covered and lashed-down Jupiter Redstone, the Chrysler technicians and the scientists from Alabama gathered in the little adobe-style restaurant.

Will Hammond led the way through the bustling diner and nodded to Debus and von Braun who were seated alone and already having breakfast. Despite the long drive, the Germans looked fresh and crisp in their navy blue suits. Hammond chose a booth near the back with a clear view of the Huntsville convoy waiting outside.

"I always like this table," Hammond explained. "We've got a great view of the road, the weather, and whatever brass might come through the door."

"What now?" Bill Hinkle asked, and Darrell was surprised with the question. If anyone knew what was next it would logically be the administrator from Chrysler. The last leg of the drive had been through the night, and mysteriously everyone but the Germans needed a shave and were tired, hungry, and road-weary.

"Now we wait for Betty Crocker," Hammond offered with his ever-ready smile. "Betty Crocker should be coming along anytime, or maybe she will meet us out at White Sands."

"Betty Crocker?" Darrell asked and looked around the Mexican diner.

"Oh," Bill Hinkle said with a yawn, "I know about that."

Before Hinkle could explain, a waitress in blue jeans and a flannel shirt appeared with an apron and a notepad. As her ponytail bounced, she stood with her hands on her hips and she was at once tough, freckled-faced, and pure West Texas. She was also quite beautiful. "Well, well, well," she said as she shook her head and smiled. "You boys look like you've had a rough night."

"That's right honey," Hammond nodded. "And now we need some Wild West coffee and huevos rancheros all around. That's a Mexican breakfast boys," the sergeant explained, "and I'll guarantee you won't be disappointed."

"That's a good one all right," the girl's accent drawled as her blue eyes danced and she wrote down the order. "But that's a meal *those two* don't seem to like." With a distasteful look and another bounce of her ponytail, the waitress threw her attention to Kurt Debus and von Braun who were absolutely consumed in an animated conversation.

After the Wild West coffee was delivered, Bill Hinkle took a sip and grimaced, "Good Lord," he said, "that's strong."

"Black Texas coffee," Hammond sampled a taste, "straight out of the chuck wagon—probably been roasting for days."

Darrell tasted his cup. "I like it," he said. "Now tell me about Betty Crocker."

"Betty Crocker goes back a long way," Bill Hinkle offered as he poured sugar into his coffee. "Years ago, back in '48, General Mills, the baked goods and cereal company, was asked by the Air Force to come up with a plan and a balloon that could lift a rocket and then launch it from about four thousand feet. This was before Redstone and before we had a dependable missile that could deliver A-bombs at medium range. The idea was it would be easier to send up a rocket if it were already off the ground. General Mills made a plastic film that was perfect for making big balloons and that's how a company that makes breakfast cereal got into the aeronautics business. That idea was scrapped, but because the same General Mills engineers were designing grain elevators, they were awarded the contract for the Redstone mobile missile launcher. It really is similar to the launching trailer the Nazis . . ." Hinkle paused and glanced over to where von Braun and Debus were having breakfast. "The trailer that the Germans," he corrected, "made to haul out their V-2s and raise those rockets to a level and vertical position for takeoff."

"Yeah, but our Betty Cocker as we call her," Hammond chimed in, "has a lot of problems. Sometimes, she stays level and straight and sometimes she don't. I've seen a lot of ballistic missiles go very wrong with the Betty Crocker recipe, and all of those that have gone south were a lot smaller than what we've got out there."

It was simple, Bill Hinkle decided. Sergeant Will Hammond was a great storyteller, and right on cue, everyone looked out through the big plate windows to observe a group of southwestern civilians talking to the Huntsville soldiers who were watching over the covered Jupiter. Even with the distance, the rocket patches on the shoulders were easily visible.

"Just like I said," Hammond lifted his coffee cup to the men outside. "Even though things look right," he warned, "a lot of times they're not. Strange things are always happening out in this desert. There are dry quicksand pits that can swallow a car, radioactive particles from the A-bombs . . . Why there's even ghost towns out there in the middle of nowhere. Whole towns built by the Army with everything modern and nice, and then blown up by A-bomb tests. You boys should see the ruins. They've got mannequins out there just like in a department store except those plastic people were melted down to almost nothing. . ."

"So you're telling me," Darrell interrupted carefully, "that General Mills, the cereal food company, makes the platform that launches the Redstone rockets."

"That's right," Hammond grinned. "Sometimes Betty stays level and straight and sometimes she don't, but General Mills was the lowest bidder on *that* contract and that's why we call the cereal company's platform Betty Crocker. And because, just as sure as God made little green apples, someone cooked that one up."

When the El Paso breakfast came, to Bill and Darrell it looked like a mess. There were roasted and browned beans mashed and spread over thin cornmeal bread with almost raw eggs spilling out over the top. On the side was a big helping of chopped tomatoes and a few deadly-looking peppers. After the Texas waitress arrived to deliver the meals, she smiled, winked, and sat a big bottle of Mexican hot sauce down on the table. "This'll fix you up boys," she drawled proudly, "one way or the other."

Sergeant Will Hammond tucked into the Mexican feast with gusto, but Bill Hinkle and Darrell looked over to the Germans and wondered what they were having.

Chapter 33

"Today, everybody remembers Galileo. How many can name the bishops and professors that refused to look through his telescope?"

— James Hogan

After the rest of the day and a starlit night at the Sombrero Motel, the Huntsville convoy saddled up and headed out into the desert. Most of the roads were perfectly straight, even through the security checkpoint where anyone who wanted it seemed could easily skirt the official entrance and find an entry anywhere along the deserted and desolate tract. In the distance, over the incredibly white sand and the sparkling gypsum dunes were the rising San Andreas Mountains. Beyond the distant mountains was Roswell, New Mexico, and the legendary location of Will Hammond's crashed flying saucer.

As promised, sixty miles from El Paso, a spanking new "Betty Crocker" launching platform was waiting beside the blast-proof blockhouses of the White Sands Proving Grounds. Stationed alongside the General Mills trailer were tanker trucks filled with liquid oxygen and methyl alcohol, more soldiers with desert fatigues and rocket patches on their shoulders, and the two Chrysler sedans Darrell had seen the day before at the Sombrero Motel.

With the help of the ever-ready soldiers, the lashings and the canvas coverings were coming off the newly arrived Jupiter Redstone as Darrell and Bill Hinkle joined Will Hammond near the exhaust cone of the recently tested engine.

The three men were standing on the desert floor and watching the rocket-patch soldiers at work.

"Not too hot today," Hammond adjusted his cap. "It'll be ninety by noon," he offered, "but it would be a hundred and fifteen if this was summer. In summer, we'd have to have shade tents over everything—even the trailers with air-conditioning. The sun out here can be a real killer."

Darrell was listening but he had climbed up and onto the tractor-trailer flatbed. He was looking inside the blackened engine cone. From the static test in Alabama, the combustion chamber was black with "shut down soot" but there were new and unblemished steering vanes attached to the inside of the exhaust cone. Outside, the combustion chamber and the fuselage of the rocket were the stabilizing fins and control surfaces that would guide the Redstone while in flight.

"Yes sir," Hammond continued, "it can get mighty hot from time to time—like back in '45 over at the Trinity site. That's only a few miles from here, and where Bobby Oppenheimer and his boys detonated President Truman's first big surprise for the Japanese. When that very first A-bomb went off, it melted a whole crater of white sand and made a big lake of shiny green glass. I heard tell, it sure was pretty—that green lake of glass—but they covered it up though—because it's still radioactive. Yes sir, there's a lot of strange and peculiar things out in this desert. Things that can kill you."

With all the lashing straps and the canvas coverings off and away from Huntsville's newest pride and joy, Betty Crocker was being pulled into a receiving position alongside Hammond's diesel tractor and flatbed. As Darrell stood on the trailer and looked down the length of the rocket, he was surprised that the upper portion of the long-range Redstone was now painted with the camouflaged Army colors of olive green, gray, and earth-tone brown.

Bill Hinkle had been preoccupied with Will Hammond's tales of the mysterious desert, and appeared mildly irritated by the narrative, but switched his attention to Darrell on the trailer. "What are you looking for?" he asked. "Is there something wrong?"

"No, no, I don't think so," Darrell jumped down from the tail section of the rocket. "It's just that those steering vanes are new — the ones inside the exhaust cone."

"Zo," Von Braun said as he approached with Debus in tow. The two men were followed by the almost always present Captain Freddy Shultz.

"I see our new guidance engineer is already in progress with his work," Von Braun said, indicating the engine cone. "You have found something perhaps — something that is out of order?"

With the southern army captain translating for Debus, von Braun waited for an answer. The Germans had abandoned their navy blue suits for the desert and now wore short-sleeve white shirts open at the neck with desert-style khaki trousers. In addition to their new uniform of the day, both of the Germans were now wearing Texas cowboy boots.

"I was wondering about the steering vanes," Darrell confessed as he pointed to the steering appendages that were spotless and soot-free. "I was wondering why they are new? Why weren't they included in the static test at Huntsville?"

"Ach, Zo," von Braun nodded as Freddy Shultz translated the question for Kurt Debus.

"The guiding vanes for early flight," the German clarified, "the initial steering controls are made mostly of carbon." He explained, "This material can withstand the heat of the combustion chamber, but the rate of the decay with the heat, the pressure, and the vibration can never be calculated perfectly. Therefore, new steerage vanes for liftoff are always installed after every engine test. We do not install the steering vanes until the prototype is ready for test flight. Otherwise, all we have is a waste of good carbon or graphite."

Darrell nodded thoughtfully and listened as the captain finished his translation for Debus. "How far away will we be during the launch?"

Von Braun gestured to the blast proof blockhouses, and then out half-a-mile into the desert where a caterpillar bulldozer was leveling a launching site. "There," he said and pointed. "You can observe the distance. This is a greater space than for our 'Old Reliable' Redstone, but I think, even with the larger fuel tanks, this will be a safe margin for error."

"I would like to be a lot closer," Darrell said, and he was amused when he heard Will Hammond whistle.

"But we have only the protection of the blockhouses," Von Braun said cautiously. "Any closer proximity without protection could not possibly . . . be safe."

"What if we made our own protection—our own blast wall at a much closer distance?"

"But how would you do that?" von Braun was incredulous. "How could this be possible? We are launching today. The weather is perfect. There can be no delay."

"That caterpillar, that bulldozer," Darrell pointed to the heavy equipment moving over the launching site. "We could pile up some sand and make a much closer observation post. We could be in position behind it. I really would like to have a good look at all the guidance systems up close and exactly at liftoff. As you said the other night, Dr. von Braun, guidance during the early flight will be critical, especially with the long-range tanks and the heavier payload. I really think we should give it a chance."

"Yes, yes, of course, I agree," von Braun nodded solemnly as Freddy Shultz was flying in German for Debus. When the captain finished translating, Kurt Debus sucked in his breath.

Before anyone could comment, Darrell said, "Of course, anyone that would want to be that close would have to be a volunteer. That goes without saying. I'll bet Sergeant Hammond and Mr. Hinkle from Chrysler will want to be up close and in front to watch. I can run a bulldozer. I learned back in Iowa. We can build a blast barrier ourselves and make it safe. We can do it in no time."

"Very well," the German agreed, "but only with volunteers. I have been a witness to too many failed experiments not to insist. You can't imagine some of our misfortunes at Peenemunde—the magnitude of those explosions."

Darrell turned to Bill Hinkle and Will Hammond, "Well boys? What'll it be? Can I count on a couple of volunteers?"

"How close are we talking?" Hammond asked with a tilt of his head toward the bulldozer.

Darrell grinned. "Close enough," he said, "to see if anything strange or peculiar happens—you know—strange things can happen out in this desert."

Chapter 34

*"There are a thousand things that can happen when you go light
a rocket engine, and only one of them is good."*

— *Tom Mueller*

With a perfectly clear New Mexico sky and the southwestern sun high overhead, Darrell and Will Hammond
took turns at the controls of the caterpillar bulldozer. For
over an hour, they had been pushing up sand and making a
blast wall about one hundred feet from where "Betty Crocker" now supported the new long-range Jupiter booster.

The camouflaged colors of green, gray, and brown were
oddly out of place as the newest Chrysler missile pointed
skyward above the endless white dunes of sparkling gypsum sand. The liquid oxygen and methyl alcohol fuel tank-
ers were in position alongside "Betty Crocker," as a team of
the Huntsville technicians strung electrical cables from the
now poised rocket to the distant blockhouses and the even
more distant generator trailers.

With the new sand wall twenty feet tall and almost complete, von Braun, Kurt Debus, Freddy Shultz, and the army
captain's aide pulled up next to the idling bulldozer in an
open jeep. Darrell stepped off the diesel machine and joined
the newcomers. Freddy Shultz, his driver, and the Germans
were all brandishing impressive-looking binoculars hanging around their necks.

Debus was speaking venomous German at a breakneck
pace and gesturing to Will Hammond who was climbing
into the operators' seat of the idling bulldozer.

Von Braun raised his hand in an effort to ward off the Teutonic onslaught and wearily shook his head. "Dr. Debus, I'm afraid," von Braun explained, "is most concerned about the position and integrity of your makeshift observation post. This is much closer than we expected. The *Herr* Doctor has pointed out that we now have a new and larger prototype that has never been tested. The fuel load is heavier than ever before, and with the heavier weight, actual flight could be delayed until a sufficient amount of fuel is burned. We must expect a delay until the fuel weight is diminished enough to permit liftoff."

"This delay—" Darrell was looking at the now vertical and suddenly menacing Redstone, "could the fuel burning delay cause damage to the carbon steering vanes?"

"Absolutely," von Braun nodded as the captain translated for Debus. "Our fuel consumption calculations are, of course, correct, however, there are always variables. And yes, liftoff could be delayed long enough to cause problems with the combustion cone steering apparatus. This is another very important reason to be cautious."

Darrell was looking at Debus, "If there *is* a problem," he said, "I really want to see it firsthand and up close. We really do need to have a look."

As Freddy Shultz completed the translation and listened to the German response, von Braun smiled. "The Herr Doctor," he said, glancing toward Debus, "thinks you are crazy, but if you insist, we will wish you the best of luck from our binoculars and the blockhouse."

With the caterpillar treads rumbling and Will Hammond pushing a precarious new pile of sand to the peak of the blast wall, von Braun looked at his wristwatch. "Are you satisfied with the protection? Or do you wish to have more time?"

"Darrell glanced over to the distance between the poised Redstone and Will Hammond on the bulldozer. "I think we're there," he said. "This should be enough."

Abruptly von Braun nodded and spoke to Freddy Shultz in English. "Begin the fueling procedure."

Chapter 35

"I think we tried very hard not to be overconfident, because when you get overconfident, that's when something snaps up and bites you."

— *Neil Armstrong*

As Darrell Loan, Will Hammond, Bill Hinkle, and two hastily acquired volunteers lay prone behind the bulldozed sand wall, the fueled Jupiter Redstone sat atop the General Mills "Betty Crocker" and waited for engine start. Chilled liquid oxygen vapor was a simmering and smoldering fog pouring out of an equalizing vent and rolling down the camouflaged colors of the rocket.

From the half-mile distance and from the edge of the blockhouse, von Braun, Debus, Freddy Shultz, and their inseparable army aides stood intently watching with binoculars. All the remote electrical connections and all the cables running from the blockhouse and leading to "Betty Crocker's" gantry tower had been thoroughly tested, including the firing range officer's "Missile Self Destruct."

Everyone knew the weather for a rocket launch could not have been more perfect. The southwestern sky was a clear and deep topaz with a light breeze out of the northwest. Visibility was virtually limitless with the San Andreas mountain range sharp and well defined.

Abruptly, a sound powered phone jingled into life beside Bill Hinkle. After picking up the army-issue handset and listening for only a few seconds, he offered the communications link from the blockhouse to Will Hammond. "It's for you," he said.

After a moment of listening, Hammond grinned and covered the mouthpiece. "It's the krauts," he whispered. "They want to know if we're ready."

Everyone laughed, including the two very young Huntsville soldiers that Hammond had convinced into becoming adventurous volunteers. Both of the young men with crew cuts were each cradling the newest German-made Leica thirty-five-millimeter cameras assigned to them by von Braun. Their instructions had been simple: take as many photographs as possible.

"Yeah, tell them were all set." After speaking, Darrell looked to the two youngsters with the German cameras. "But listen," he warned, "when this thing starts we won't be able to hear or to talk, and if it goes bad, we'll all need to get down to the bottom of this sand pile real quick. Everybody understand?"

Everyone nodded somberly.

"Boys let me tell you," Hammond nodded in agreement—he was still holding the muffled handset, "I've seen enough to know. If I start for the bottom of this hill, you had all better be moving with me."

"Ok, that's it," Darrel agreed. "If one of us starts down the hill, we all go for cover together."

"Hell in a hand basket," Bill Hinkle offered, "I can't believe we're this close. Tell the krauts to get started and let's get this over with."

Hammond nodded and spoke into the sound-powered phone, "Forward observation ready. Repeat. We are ready when you are."

Even before Hammond could replace the handset, the accelerating whine of the turbine fuel pumps was audible.

Within seconds, the now familiar heat flash jumped across the sand and covered the top of the makeshift dune. The thundering combustion that followed seemed to shake and rattle "Betty Crocker" until the launching platform and the ignited rocket were only a pulsating blur.

The two rocket-patch soldiers with the cameras were clicking away.

Even though the new Jupiter began to throttle up, she still sat stubbornly in place with the blazing exhaust burning, pounding, and consuming the General Mills platform with every passing second. Just as it seemed there would be no launch, but only a destructive blowtorch cooking Betty Crocker into ashes, the largest Redstone ever began to climb. As the painfully slow-rising rocket cleared the gantry tower, and as the fiery engine combustion rose above the volunteers on the observation dune, the first signals of trouble began in the form of unnatural debris falling through the slipstream of the expanding exhaust. At first, it was only minuscule particles flashing white-hot in the detonating blue flame, but then larger pieces began to fall as the lifting Redstone began to tilt dangerously.

When Darrell looked away from the disaster that he knew was about to happen, he felt Will Hammond clap him on the shoulder. There was no mistaking the signal and the action, and as a group, all five of the forward observers tumbled down the sand for the refuge at the base of the makeshift dune. When the men were at the bottom of the sand hill, however, the failing booster was high enough for all to see as it tilted and rolled more dramatically, and then began to tumble out of control.

Before anyone could consider the firing range officer and the inevitable self-destruct command, the failing Chrysler missile angled over and gathered speed. With the acceleration incredible, the still-thundering engine propelled the slender column of volatile fuel tanks into a downward angle and then into the earth. To the volunteers behind the sand wall, the shattering detonation felt like an earthquake. The Jupiter explosion was only fifty yards away.

Meanwhile, at the blockhouse and the command post, the Germans with the binoculars were over a mile away from the explosion, the new crater, and the black smoke that was suddenly rising over the desert. The column of smoke was black, the sand very white, and the perfectly clear New Mexico sky tainted with the latest lesson in rocketry.

When the concussion wave of the blast was over, sirens began wailing from behind the blockhouse and three jeeps full of soldiers began barreling for the bulldozed dune.

Without speaking, all five of the forward observers climbed back up through the sand to their original position on top of the dune—except now they were standing.

"Mother of God," Bill Hinkle exclaimed as he looked at the black smoke that was rising. "I never want to go through that again," he declared. "We were almost cooked, and now Chrysler is going to have a fit!"

"I told you," Will Hammond offered sagely as he dusted off his fatigues, "Strange things can happen out here; mighty strange. One minute the desert's normal and the next crazy."

"That wasn't strange, or crazy," Darrell said as he stood atop the mound and watched the still burning crater. "That was predictable. Did you see those steering vanes?" he asked. "Did you see them coming apart and falling away? No wonder she couldn't fly. She lost her direction."

Hammond grunted. "That don't explain why the self-destruct didn't work. Tell me about that!"

"I'll bet we owe that one to von Braun or maybe even Debus," Darrell turned to the approaching jeeps. "If that range officer had pushed the self-destruct, we probably wouldn't be here to talk about it. At the first sign of trouble, our girl was lifting right over our heads; if they had signaled that self-destruct—we would all be gone—no doubt about it. No one could have lived through a bath of that burning fuel and oxidizer."

After Darrell spoke, the older men focused on the two young soldiers that were Hammond's volunteer photographers. One of the young men appeared to be in shock, his face gray, without any color, and the other was on his knees, getting sick in the sand. Both, however, were still clutching their cameras.

When the jeeps arrived at the base of the dune, Hammond nervously lit a Lucky Strike. "Smoke 'em if you've got 'em boys," he said. "I think we deserve it."

Von Braun and Debus were out of the leading jeep instantly and climbing the hill. "Is everyone all right?" Von

Braun was observing the two shaken soldiers holding the cameras.

"We're okay Boss," Hammond said to the German. "A little shook up I suppose, but I guess we'll manage."

"*Mein Gott!* Thank God for that," von Braun clasped his hands together. In the next breath he asked, "Did you see what happened? Could you observe the malfunction?"

"We saw it all right," Darrell nodded. "The carbon steering vanes just fell apart. It was like they melted away."

"That's right," Hammond chimed in. "We saw everything, and I, for one, will never forget it. That was way too close for comfort and mighty strange."

Chapter 36

"The man with the new idea is a crank . . . until the idea succeeds."

— Mark Twain

"Tell me everything," Von Braun said, "omitting nothing. I would like to hear from everyone individually."

Will Hammond, Darrell Loan, and Bill Hinkle were seated inside the largest blockhouse along with the two young soldiers who were the last minute volunteer photographers. Freddy Shultz was waiting in the background along with his aides and all of the contributing technicians that had been watching the test flight from the blockhouses.

The desert air was cool at sunset, the temperature pleasant, but there was still the scent of the smoldering rocket crater from over a mile away. During the Jupiter remains clean up and the examination of a severely damaged "Betty Crocker," the film from the forward observation post had been developed. There were now eighteen large and glossy photographs spread out under the lights.

Von Braun, Debus, Freddy Shultz, and his crew were standing and waiting patiently and clearly very upset — but no one more so than Bill Hinkle. The new Jupiter failure was a total disaster and the upcoming consequences with Chrysler, the military, and the brass in Washington were obviously weighing heavily. An Army failure so similar to the Navy's Vanguard launching explosion was something no one wanted to consider.

"Well," Hammond began, and the blockhouse went silent. "I guess I can go first," he offered. "I never saw a lift-off like that one. It took too long. It was like the old Jupiter was doomed from the start. I knew right away we should have all gone down the hill because it sure didn't look right, but something held me there — I just had to stay — it was like watching a bad wreck."

When the sergeant finished he stood away from the chair von Braun had placed. Darrell rose, as did Bill Hinkle, and the two youthful photographers. The situation was too uncomfortable for anyone to sit. Everyone was now standing around the table and looking at the glossy photographs. Hammond moved past Debus and pointed to a black-and-white photo that was taken just before the test flight tilted out of control.

"Right there," Hammond tapped the glossy. "You can see it in the exhaust — all the steering gear coming apart. See the white flashes?"

"*Ja, Ja*, of course," von Braun said testily, "but was there anything else — perhaps an apparent cause for the malfunction and the failure?"

Bill Hinkle shook his head, "No, never seen anything like it," he said. "And I've seen dozens of photographs from dozens of launches and I've never seen an 'Old Reliable' Redstone do anything like that. It looks like we might be going back to the drawing board. This is something Chrysler is not going to like."

Abruptly, Debus exploded in German after he heard Freddy Shultz's translation and the blockhouse erupted in multiple conversations.

After a moment, von Braun raised his hands for quiet and spoke with authority, "We are *not* going back to the drawing board because of faulty steering vanes. We must correct the problem and then move on. Did anyone else see anything?"

"I believe we all saw the same thing," Darrell said, "and it's visible in these glossies. We need steering vanes that can take more heat, and take it longer."

"Of course this is logical," von Braun was concentrating on the photos depicting the falling debris, "but nothing we have found can withstand the violent temperature except pure carbon."

"I think the problem is not the carbon," Darrell spoke thoughtfully. "The problem is," he paused, "that the carbon is pure." After he voiced his thoughts, the newest Chrysler troubleshooter pulled a pen out of Will Hammond's shirt pocket, flipped over one of the glossy photographs, and sat down again in one of the chairs.

"It's the pure carbon that failed," Darrell said as he tapped the pen on the back of the photo. "What if we make new steering vanes of stainless steel and then impregnate the metal somehow with carbon." Darrell began drawing and everyone gathered to look over his shoulder. Every soldier and civilian in the blockhouse moved closer.

Patiently, and with a clear hand, vertical lines began to appear illustrating sandwiched layers gathered into the shape of the original carbon steering vane: a white space, then a dark space filled with ink, and then more white and dark spaces in succession.

Suddenly Debus sucked in his breath, flipped over another photo, and began his own drawing with a hastily pulled mechanical pencil. After a moment, his illustration showed a three-dimensional depiction of Darrell's sandwiched materials but with holes obvious in the sandwiched vanes indicating carbon layers running in two separate directions. The sandwiched stainless and carbon idea was the same, but with carbon rods running across horizontally as well as vertically up and down. There was a system to Debus' adaptation that was entirely three dimensional.

"*Mein Gott*," Von Braun began in German and then switched to English. "That's it!" he exclaimed proudly. "The hardened steel will hold the carbon together, and when the outer portions burn away there will still be more of the same material and more of the same layers deeper inside. Excellent—brilliant! Absolutely brilliant!" he said again.

At once, Debus' flowing German was unstoppable, and even with the language barrier, it was obvious that he was praising the newest appointee from Chrysler for the inspired insight.

With everyone encouraged and beginning to crack smiles, especially the two young soldiers that had been Hammond's up-close-and-personal volunteers, von Braun reclaimed all attention. "We can test the new impregnated steering vanes during preflight engine tests at Huntsville," he declared with a nod, "and learn just how long we can depend on the new material."

"Congratulations, Mr. Darrell Loan," von Braun offered expansively, "there can be no question you have ignited the spark of genius!"

"Does this mean we already have a new design?" Bill Hinkle asked carefully. Clearly, he was worried. "A new design to replace the failed problem with our newest and most expensive booster?" he said. "Something I can tell Chrysler about before they go through the roof about what happened today?"

"I feel certain," von Braun nodded confidently, "that we definitively have solved the problem with a new stabilization design, and we will now be unstoppable in our race against the Communists. This is wonderful, *ja, ja,* simply wonderful."

Chapter 37

"Our prime obligation to ourselves is to make the unknown known. We are on a journey to keep an appointment with whatever we are."

— Gene Rodenberry

With the new guidance concept tucked away and headed for Huntsville, Bill Hinkle and Darrell flew home to Detroit. When they arrived in Michigan, another new Chrysler was waiting at the airport. The evening sky was dark, it was heavily overcast, and the December snowflakes were falling and coming down in the headlights. During the drive to Warren, and to where Audrey was waiting, Darrell saw how the failure in New Mexico had taken a toll on the administrator from Chrysler. Hinkle was depressed but trying to be helpful.

"I hope everything will be all right with Audrey," the Chrysler man offered as the heated car drove through the falling snow.

"She'll be all right," Darrell spoke over the singing tire chains and the announcer on the radio, "but what about you?"

"As you know," Hinkle said, "I'm not married to anything but Chrysler, but right now, *my wife* is on the warpath. Everyone was so confident that the new Jupiter would fly without a hitch. After all," Hinkle shook his head, "the White House was watching this one and the pressure that Ike is putting on headquarters is hard to imagine. To stay out of trouble with Washington, we've promised a good test flight before the New Year and we're working around the clock to make that promise happen."

"Don't worry too much," Darrell offered over the sounds of "Silver Bells" beginning on the radio. "If I know the Germans, they won't rest until the new steering vanes are perfect. I really believe that's the only problem, and the next flight *will* go off without a hitch."

Before Bill could comment, he turned off the road and into the drive of the new home in Warren. Both the arrivals from White Sands were surprised when they saw three Chrysler sedans parked in the driveway. All the lights were on in the house and on the porch. There were Christmas lights glowing from under the snowfall and not just from the neighbor's decorated spruce but from the new holiday decorations Audrey must have set up. There were colorful lights around the front windows, around two snow-covered shrubs on either side of the walkway, and from electric Christmas candles glowing red on either side of the front door.

Before Bill Hinkle could do anything but pull in behind the last Chrysler in line, Darrell said, "Come on in Bill. You might as well help me out here, and besides, I think you should do something else other than go home alone and worry about what Ike is doing in the White House."

After Darrell and Bill tracked through the snow and were cleaning their feet on a new doormat, they could hear Bing Crosby singing "Holiday Inn" from inside the house. When the door opened, at once it was obvious the new home was feminine filled, festive, and packed with The Chrysler Wives Club.

With Darrell and Bill Hinkle stop-gapped and gaping in the foyer, Audrey swept across the floor with a dangerous smile. Her eyes were mischievous and hair was perfect. She was wearing a green dress with a red poinsettia pinned to her lapel, as were all the other Chrysler wives, each in their own holiday finery.

"Welcome home," Audrey said as she pecked Darrell on the cheek. "After you called, I thought we might as well have a little welcome home party. Either that or a fight, but I decided a party would probably be better."

When Bill flushed with embarrassment, Audrey placed her hands on her hips. "You too Mr. Bill Hinkle," she said, and her smile was still unsafe. "We could be in a fight too after you dragged my husband away after only one night. But now," Audrey glanced to the other women that were carefully watching but pretending not to be, "we can all be friends again especially after hearing about you being all alone for the holidays and without someone special."

"Sally," Audrey called for the pretty redhead, but the beauty with the auburn hair was already waiting to advance and she was not alone.

"Bill," Audrey said, "I believe you know Sally Booth, but I don't think you know her little sister Darcy."

After drinks and sandwiches were served, along with Christmas cookies, brownies, and coffee, Darrell obediently tended the fireplace. After new logs were placed on the fire, Audrey sat on a corner of the couch and contentedly looked around the room. The party was a success she decided, and much better than a fight, and now with the wives club packing up and heading into the snowy night she was very pleased she had done all she could for Sally Booth, her little sister Darcy, and Bill Hinkle who was obviously a catch, but also clearly a stubborn bachelor.

Audrey decided men would always be a difficult study, but at least her man was home for Christmas and it was about time. Thank God for the wives club she thought, and for finally understanding that this new life in Michigan would be much better than New York. With the Chrysler network of women, anything could be possible.

Sally Booth and Darcy were the last girls to leave, and after Bill Hinkle said good night, Darrell and Audrey were alone in the firelight. Darrell poured a nightcap and sat beside the fire. He noticed Audrey watching intently. Her gaze was thoughtful but pleasant.

"All right," she said, "tell me why you look so worried?"

"It's not really that I'm worried," he confessed, "but we had a failure on the launching pad—almost like what we saw on T.V. with the Navy. You remember the Vanguard missile, the one that blew up in Cape Canaveral?"

"I know about that," Audrey said carefully. "All the wives know about what happened in New Mexico, but what I want to know is what's next?"

Darrell sipped his drink, "The plant at 16-mile is working overtime to make more prototypes just in case we have another problem. The budget must be incredible because the White House is pushing so hard."

"I know that too," Audrey smiled. "This is not like New York. The wives here in Michigan know pretty much everything."

"Well that's good," Darrell grinned. "At least I won't have to break any rules about top secret clearance if you already know what I'm doing. Hey Audrey," he said, "are you sure you don't want a nightcap? I didn't see you drinking anything all night."

"That's right silly," Audrey smiled again. She was very pleased with Michigan, Christmas time, and the Midwest. "Not much drinking for me," she announced happily. "We're going to have a baby."

Chapter 38

"Motivation is the art of getting people to do what you want them to do because they want to do it."

— *Dwight Eisenhower*

Bill Hinkle sent flowers for Audrey on Christmas Eve because Sally Booth and her little sister Darcy had already told him about the baby, but on the day after Christmas, Darrell and Bill were away again and flying down to New Mexico.

The airport was clear and cold in Michigan with sunshine so bright on the snow it was hard on the eyes, but after the two men from Chrysler changed planes in St. Louis, the bumpy ride to El Paso was filled with clouds and violent turbulence. The wind was gusting when the Douglas DC-3 touched down in Texas, and before the airliner could taxi to the waiting cars at the terminal, one of Will Hammond's fabled sandstorms was a howling monster moving over the desert.

The mounting pressure from Chrysler was almost a visible yoke on Bill Hinkle's shoulders and the sandstorm arriving the day before the next scheduled Jupiter test was making matters worse. As two youthful soldiers with the familiar rocket patches hastily loaded the new arrival's luggage into a waiting car, they were obviously concerned about the severity of the upcoming storm. Bill Hinkle was concerned about everything.

"It's a good thing you landed when you did," the driver from the motor pool explained nervously. "The tower just closed the airport. You guys were the last plane in."

136

With the car doors closed against the darkening threat, the gusts began, the swirling desert moved over the airfield, and sand crossing the runway began blasting the windshield. Over near the terminal, the pilots and the airport ground crew were struggling under the wings of the airliner to tie the plane down to the tarmac.

During the slow and tedious drive toward the Sombrero Motel, visibility in the fading afternoon became almost nonexistent and the growing tension in the closed sedan worrisome. Even with the headlights on bright and the windshield wipers clawing at waves of increasing sand, there was nothing visible but a few feet of cracked pavement ahead and the occasional pair of red taillights crawling forward in the distance. When at last the buffeted Chrysler pulled up to the adobe-style restaurant and the flickering neon sign glowing dim over the Sombrero Motel, the wind and the relentless sand particles were gusting to over fifty miles an hour. Positioned in front of the plate glass windows and very near the restaurant another tractor-trailer from Huntsville waited with a new Jupiter Redstone lashed under canvas tarps.

With the wind and the howling desert reaching a new fevered pitch, the seasoned driver and his partner in the front seat looked to each other in concern. With the storm peaking, the soldier on the passenger side turned to the Chrysler men in the back. "Let's forget the luggage for now," he suggested, "and just get inside. We need to get out of the wind. If this is your first sandstorm," he warned with a grin, "try not to breathe too much."

When all the doors opened at once, the desert was everywhere, the sky dark, and the wind tearing. The ever-fine particles of gypsum were blasting and a part of the air. Although it was only a short distance to the motel entrance and to the restaurant's shelter, it was very clear why the soldiers had been so concerned. Will Hammond was right. Catching a breath in a sandstorm was not easy, and it was also suddenly obvious that Hammond's other warning about stalled vehicles with sand-clogged carburetors was also a likely possibility. Cars could stall and people could become trapped and panic. It was easy to imagine a lost motorist and a vehicle abandoned in a storm. The motor pool driver and his escorting partner's noticeable relief to be inside the adobe structure and out of the wind brought another worry forward and Bill Hinkle frowned and shook his head.

"This can't be good," he said as he looked through the sandblasted windows. The tractor and flatbed with the covered rocket were only dimly visible. "Who knows what's going on with this sand under those tarps? This stuff is so fine it's probably up and into the engine cone, the combustion chamber, and who knows where else. Particle contamination is why the Chrysler plant at 16-mile is so clean—one tiny fragment of anything in the wrong place and God only knows what could happen."

"Speaking of God and wondering what could happen . . ." Darrell was tired of the contagious stress, Hinkle's worries, and he was ready to change the subject, "Will you look and listen to that?"

After he spoke, Darrell decided the friendly face of the man from Chrysler was, after the latest Redstone disaster, only a memory. Across the darkened restaurant and an expanse of empty tables and chairs, the lights were on in the adjacent cantina-bar and the sounds of piano and German voices in song were audible above the gusting wind and the desert slashing at the windows.

The rocket-patch soldiers that were the driver and the airport escort had removed their caps and were dusting off the storm-driven sand. They were waiting patiently in the foyer as seasoned veterans of worry and concern, but after a moment of silence for Bill Hinkle's troubled comments, the soldier that had been the driver tilted his head toward the lights and the drifting German music.

"Mr. Hinkle," the soldier held his cap in his hand, "don't worry too much about the sand. We get these storms a lot. And the Boss," the young man suddenly smiled, "he knows just what to do. And besides, it's sing-along-night with beer and everyone's invited."

"Sounds like it's only for the Germans," Hinkle's tone was as dejected as his mood.

Before Darrell could insist on having a look, Will Hammond appeared and sailed through the empty restaurant with a bottle of beer in his hand.

"Well, well boys," he said beaming, "I thought I heard the storm get a little louder. Come on in — what are you waiting for? We've already started. After all, it's kraut night," he whispered and winked, "and the Boss is in a good mood. He's buying the beer. "

"Come on Bill," Darrell urged, "let's see what's going on."

With the driver and the escort leading into the bar, Hammond lowered his voice and said, "They sound just like the Nazis in the newsreels. You'll swear they're singing the old Adolph marching songs."

It was true, the Germans did sound like something right out of the war, and with the dark storm raging outside, the bomb shelter effect was perfect. Von Braun was standing in the dim lighting by the piano that Debus was playing, and Freddy Shultz was standing by von Braun and several of the other rocket-patch soldiers. Just as a raucous and guttural marching song ended, all the men joined in with Debus on the piano for the last verse of "Lili Marlene." The previous song was in German, but now everyone was singing in English and the voices sounded like a well-rehearsed choir.

Resting in a billet . . . just behind the line
Even though we're parted . . . your lips are close to mine
You wait where that lantern softly gleams.
Your sweet face seems to haunt my dreams
My Lilli of the lamplight . . . my Lilli Marlene

When the verse was finished, von Braun noticed the newcomers and motioned them over.

"Gentlemen, gentlemen," the German accent offered warmly, "welcome home! Bartender," Von Braun gestured to a Texas civilian behind the bar, "beer for our latest guests!"

When the beer came, von Braun smiled and offered a toast: "*Prosit*, and to your good health!" he said as he clinked beer glasses.

"We have done it," the German declared simply and then sipped his beer. After the statement, Bill Hinkle looked hopeful, expectant, and worried all at the same time.

"We have perfected the new carbon and steel steering apparatus," Von Braun explained confidently, "and the static tests in Huntsville were more than satisfactory— better than we could have hoped. The new composite material lasted four times longer than the estimated time required."

"Then you're confident about the next test," Bill asked carefully, "even with all this sand?"

"Absolutely," Von Braun nodded and then turned to the men by the piano as Debus began pounding out another wartime German classic.

"*Ach!*" Von Braun exclaimed as he clearly was referring to Kurt Debus. "The man is a genius! Come join us by the piano—I insist."

Chapter 39

"I believe every human has a finite number of heartbeats. I don't intend to waste any of mine exercising."

— *Neil Armstrong*

The sandstorm raged through most of the night, but when the sun rose over the desert, the air was clean and clear with a bright blue sky and the late December temperature in New Mexico was perfect.

It was only seventy degrees at noon when Will Hammond and the rocket-patch soldiers finished hauling the newest prototype out to the proving ground and positioned the latest Jupiter atop another newly arrived Betty Crocker. As all the soldiers, scientists, and engineers worked toward their individual goals required to fuel and launch a Chrysler missile, an air of quiet determination was apparent in every focused detail of every assignment. All of the personnel were well aware that another failure would be dangerous for "The Boss" and horrific for the Huntsville crew and the US Army.

Once again, the gantry crane had lifted and poised the waiting Jupiter, the tanker trucks had fueled the volatile long-range tanks to redline capacity, and the telemetry and power cables were in place and trailing from the blockhouse. For the latest test flight, however, Bill Hinkle was now stationed in one of the blast-proof bunkers and waiting with the Germans, as all but the forward observers were watching with binoculars from the half-mile distant blockhouse.

The original forward observation sand hill had been blown away by the storm and now blended with the other wind drifted dunes. At von Braun's urgent insistence, the newest protection wall of bulldozed desert was three times as far from the launch site as the first makeshift wall of sand. With the new plan in place, there were four forward observers waiting about one hundred yards from where the fueled Jupiter was waiting. It was clear that the rocket was ready. Venting liquid oxygen was a pouring vapor running down the fuselage and gathering as fog at the base of the launching platform.

Will Hammond, Darrell, and the two rocket-patch soldiers that were the volunteer photographers on the previous failed launch, were ready again with the Leica cameras and fresh film. Once again, the forward observers were waiting for engine start beside the sound-powered phones and the only communications link with the blockhouse. As harrowing and as violent as the first failed Jupiter had been, the two youthful soldiers enjoyed a certain amount of bravado from their comrades over the frontline experience. They were now excited with the prospect of another up-close adventure to elevate their popularity rank and elevation status higher. Hammond was already a legend and a troop leader for envious adventure, and his newest compatriot, the tall cowboy-style engineer from Iowa that had the audacity to suggest the forward observation post from the onset, now held a rank of honor that was only whispered about. Among the rocket-patch soldiers and most of the scientists, Darrell Loan was now: *the Cowboy*.

When the sound powered phone jingled into life, Hammond picked up the handset and answered with a simple, "Hello."

He then listened briefly and looked to the soldiers with the cameras. After both of the young men nodded that they were ready, Hammond said, "Thanks and we're all set." He replaced the handset.

"That was worried old Bill Hinkle from Chrysler," Hammond said with a lopsided grin, "Everybody at the blockhouse wishes us luck — better luck than last time."

Darrell couldn't help but smile. And then as one, all four of the forward observers heard the fuel pump whine begin in the distance. Because Debus wanted a more distinct view of the Jupiter, the new rocket was scored with black diagonal stripes in order that the pitch, yaw, and roll movements could be observed more accurately.

Just as before with engine start, the heat flash of ignition jumped across the sand, but no one could ever really be ready for the wall of sound and earth-shaking vibration that followed. Everyone instinctually ducked down behind the makeshift wall, but as the sound wave began to thunder, it was clear that the latest prototype was behaving differently. The new Jupiter was already rising.

Almost before the photographers could react and begin their documentation, the newest Chrysler achievement was well underway and climbing rapidly. The blue flame exhaust was already clear of Betty Crocker, and as the critical and early seconds of the rocket flight were passed, it was obvious that the refractory steering vanes were operating perfectly.

The missile rose steadily as Darrell, Will Hammond, and the soldiers with the clicking cameras now stood to watch. It was obvious there was no further need for the protective wall of sand. This new Jupiter was up and away and accelerating fast. It was a beautiful and moving sight and as the wall of thundering sound began to diminish, Will Hammond found his voice and began shouting in the rumbling wake, "Go! Go! Go!"

"Wow look at that," Hammond shouted over the rippling but still pounding exhaust wave. He was clearly delighted and his mood was contagious.

"You did it Darrell," he shouted over the sound. "Your idea worked!"

Darrell was beaming. "It was just as much Debus as it was me," he offered over the rumble.

"Bullshit," Hammond growled back. "The kraut took your idea and ran with it like a touchdown pass. But you gave him the idea and there's no doubt about it."

"That's right," the closest photographer chimed in, his

face youthful and encouraging. "We were all there. We saw it and we all know how *that* happened."

Darrell didn't comment. He didn't have anything else to say. He was very content to stand on the bulldozed dune and watch the almost out of sight exhaust flame and contrail of smoke as the sound wave continued to diminish. *This is the real beginning,* he thought. *The start of what we all really need.*

Before there was a chance for further comments or more congratulations, the distant rumble of combustion died away and was replaced with a sudden double-boom of a thundering and violent explosion. There could be no denial — the latest Jupiter Redstone had just exploded.

"Damn!" Will Hammond muttered as he kicked the sand. No other words were spoken or could be appropriate after such a spectacular beginning, but after the moment the double-boom echo drifted away, the forward observers soundlessly began walking down the dune and toward the distant huddle of blockhouses.

Chapter 40

"Don't tell me that man doesn't belong out there. Man belongs wherever he wants to go — and he'll do plenty well when he gets there."

— *Wernher Von Braun*

Outside the largest blockhouse, the short winter day was fading into sunset and Wernher von Braun was holding court. He was resplendent in his desert uniform of the day with open white shirt, hanging binoculars, and khaki trousers, and when he placed a cowboy boot on the running board of an Army jeep and climbed into a speaking position, his boyish good looks and winning smile had never been more pronounced.

"Everyone please. May we have some quiet?" he asked with his familiar inflection. "I know there must be many questions, but I can assure all that are present that there is only one answer."

After the earth shaking double-boom of the Jupiter explosion, all the technicians from all across the Proving Grounds gathered outside the command blockhouse. With the mood among the rocket patch soldiers gloomy at best, and with everyone speaking at once and speculating on what might have happened or gone wrong, Freddy Shultz emerged from the fortified White Sands structure and asked that everyone please wait quietly.

"Just give us a few minutes," Freddy's southern accent pleaded. "It's not bad, not like you think, and the Boss will be out shortly to tell you all what happened."

As promised, with every technician and soldier very concerned, and with speculating and worried conversation again beginning to mount, von Braun reclaimed all attention as he waited for everyone remaining in the blockhouse to come outside.

"Our newest long range Redstone booster that we have now named the Jupiter-C," the German announced, "has made a successful test flight and all telemetry results show a flawless performance. Our test today was a complete success."

With the sudden volume of rising voices overwhelming von Braun's efforts, Freddy Shultz climbed up on the other side of the jeep and shouted for quiet. "Now damn it!" the southern accent demanded. "I've asked for quiet and I want everyone to shut up and listen!"

"*Ja, Ja,* please some quiet. I know you all, and I know everyone here is more than concerned, and I must apologize for not coming out and making this explanation sooner. But I must clarify; I too was caught unaware, and was at first very upset with the actions that have happened here today."

With the new silence complete von Braun continued, "As I have said, the new Jupiter-C has performed perfectly. In fact, our latest achievement has reached beyond the boundaries that have been set forth by the military. At precisely sixty seconds into the flight, the Jupiter-C test was terminated when radar calculations indicated the flight path was over fifty miles above the earth. A self-destruct signal was transmitted to that altitude, and this function also operated perfectly. This is the reason for the explosion," von Braun paused and held up his hands as a new wave of conversations erupted over the assembled men.

"Doctor von Braun, if I may," Freddy Shultz interrupted. He was once again visible above the crowd and clearly demanding attention from his elevated position on the jeep. "I would like to explain," the captain began patiently with a forceful voice, "that there have always been standing orders that upon success, and upon reaching an attitude of fifty miles high, all ballistic missile tests are to be terminated at that specific altitude. I repeat, the standing orders stipulate

146

that the U.S. Army is not allowed to experiment with a ballistic missile any higher than fifty miles high. This is the reason the Jupiter-C was detonated today."

Will Hammond was a tall figure standing among the others and beside Darrell and Bill Hinkle and the two young soldiers with the cameras. "That's the craziest thing I ever heard of," Hammond said quietly from the corner of his mouth. "That's just plain crazy."

"Well," Darrell said softly, his good humor and nonchalance restored with the explanation, "crazy things can happen out in the desert."

"You want to talk about crazy, "Bill Hinkle offered quietly, "you should have seen von Braun and Debus in the blockhouse when Freddy Shultz ordered the range officer to hit that self-destruct. I thought World War II was going to start all over again!"

Chapter 41

"Farming looks mighty easy when your plow is a pencil and you're a thousand miles from the corn field."

— Dwight Eisenhower

In the blockhouse just after the launch when it became clear the Jupiter was up and away and rising quickly, Debus dropped his binoculars and ran over to read the telemetry information that the rising rocket was transmitting. His excited German was unstoppable as he rattled off the fuel pressure levels, temperature, gimbals and gyro angles, and as the Jupiter climbed higher and faster, his excitement levels were equally off the charts until Freddy Shultz crossed the room with his unheard-of orders for the self-destruct command.

Of course, Debus at first didn't understand the English instructions, but when von Braun dropped his binoculars in surprise and turned on the senior army officer outraged, both the Germans were relentless in their efforts to stop the self-destruct. The firing-range officer had already been well briefed and transmitted the signal immediately after being ordered, but the German onslaught that followed the detonation was furious.

When Freddy Shultz finally began to regain control over all of the technicians and the waiting engineers, the men in the blockhouse began their protests in earnest at the waste of a perfectly good launch vehicle and a runaway success. As every veteran of White Sands knew, almost all of the booster failures happened within the first thirty seconds after engine start. To waste a highflying booster that was already well past the initial danger point was a shame and a waste of taxpayers' money.

Most of the dialog in the blockhouse had been confrontational, and as explosive as the thundering self-destruct. Tempers flared but were professionally squelched, sides were taken that were not to be forgotten or easily forgiven, but after tumultuous argument and a tentative standoff, everyone realized that all the personnel from all over the Proving Grounds deserved an honest explanation. The rocket patch soldiers and the technicians waiting outside desperately needed answers and they needed to know that second Jupiter mission had been a success — even though, the possibilities of touching the edge of space had been denied by a rubberstamp bureaucracy a thousand miles away.

Chapter 42

"I like to think that the moon is there . . . even if I'm not looking at it."

— Albert Einstein

With the desert sunset only a fading glow in the west and the Proving Grounds crews wrapping up the final details of the day, mixed emotions washed over the rocket-patch soldiers and the Huntsville technicians as a full moon rose over the mountains in the east. Tensions were still running high between the military, and most of the civilians and gathered huddles of determined individuals were silhouettes in the dusky twilight. As the lunar body climbed higher, the darkened figures loomed over the sand and began to cast moon shadows.

A few stars were beginning to shine overhead as Darrell and Bill Hinkle stood quietly smoking with Will Hammond. When Wernher von Braun approached in the moonlight, Mars was a glittering beacon over the western mountains, but the full moon rising easily dominated the early evening. With the white sand reflecting the light from the moon and the desert free of even the slightest breeze or a hint of the sandstorm from the day before, the sculpted features on the lunar face were incredible and mesmerizing.

"Zo, gentlemen," von Braun said as he approached, breaking the silence, "may I join you?"

"Sure thing Boss," Hammond offered quickly and ground his cigarette into the sand, "we're just taking it all in — everything that's happened."

150

"Yes, of course," von Braun nodded, "a disappointment to be sure, but only one of many, and still we go forward, however slowly."

Most of the rocket-patch soldiers were packing up and loading into the waiting transports as the four men stood waiting and looking at the moon. Von Braun still carried his binoculars and lifted them easily to peer at the lunar surface.

"I have dreamed of this for almost thirty years," he explained as he looked through the lenses, "an exploration and a spaceflight mission to the moon. Even during the war and at a time when we were pressed to increase our rocketry research further, even for the efforts of destruction, my hopes were secretly for a manned spaceflight venture that could reach the moon."

The German sighed, "Even during a time of chaos," he confessed, "and during a time when we all knew the war was lost . . . I was still dreaming."

In the distance, Freddy Shultz and Kurt Debus were still obviously at odds and still arguing in German, but after a moment, Shultz opened the door on one of the Chrysler sedans. Debus ducked into the back and Freddy climbed into the front passenger seat. As the car began to move out and bounce over the shallow sandbanks, the headlights came on and von Braun continued quietly. He was apparently unconcerned or oblivious that his transportation back to El Paso had driven away without him.

"Ja Ja, you must understand," the German explained as a conspirator, "there were secret plans for a much larger V-2 — it was to be called the A-11. This was a very heavy booster with multiple engines and a prototype that could have been launched from Germany to certainly reach New York. Hitler approved this design so that an attack on the United States was possible. But I, most secretly, wanted this largest and most ambitious design ever for the moon, or perhaps for Mars or the planets beyond."

Abruptly, von Braun lowered his binoculars and handed them to the guidance engineer the rocket-patch soldiers had privately nicknamed the Cowboy.

As Darrell lifted the powerful lenses, he was not only impressed with the optics but with the clarity and the details on the lunar surface. Huge craters and mountain ranges were visible alongside vast desolate tracks that appeared to be as deserts or sandy oceans and not unlike the dunes and the terrain that were the White Sands Proving Ground.

"If completed," von Braun sighed with his eyes still on the moon, "our A-11 would have been the first creation ever that could have reached past the grip of gravity and escaped the pull of the earth. The A-11 as designed could have made a passage to the moon."

After moment of silence, Darrell offered the binoculars to Bill Hinkle, and after he too began looking aloft and focused, von Braun continued softly. It was clear the Boss was unloading his frustration and the stress of the day.

"My experience in rocketry has unfortunately always been the same. The funding comes from the government, the military regulates the funding, and when an era of breakthrough begins, a bureaucrat with no imagination takes control and we are forced to take steps backward. But now, there is a new urgency and an open competition—a contest of nations, but also I think, a contest for spies. My original A-11 was a design with multiple engines, and now our Russian counterparts have a booster with multiple engines."

Without warning, von Braun fell into an ominous silence and stood watching and waiting for a reaction from Bill Hinkle. When the Chrysler man became aware of the scrutiny, he was suddenly stoic and on guard. Without a word, he passed the binoculars to Will Hammond. At once, the Huntsville favorite lifted the lenses and focused aloft.

"Of course, Mr. Hinkle knows of the new urgency and my promise to orbit a satellite before the second week in February," von Braun's accent became more pronounced as he spoke faster. "But if we had been allowed today, and if the final stages had been in place to lift our achievement higher, I believe we could have already accomplished our mission and would be on our way to the ultimate goal of a manned spaceflight mission and a trip to the moon."

"Oh yes," von Braun added, as Will Hammond lowered the binoculars with disbelief, the sergeant's doubt and skepticism clear and obvious in the moonlight.

"Even now," the German smiled knowingly, "there can be no doubt. The next plan for our Russian Communists is a conquest of the moon. There can be absolutely no question we will very soon be engaged in the most significant contest ever for mankind — I have hoped for this — and now that this strategy has been confirmed — from the highest authority — I could not be more pleased."

Before any denial was possible, all attention turned toward two sweeping flashlight beams and two moonlit figures in rumpled suits that carried the probing flashlights.

"*Ach*, zo, my watchdogs," Von Braun said testily as the two men steadily approached. "They follow me everywhere," he explained, and his frustration inflected his accent even more. "They listen to all my telephone conversations. They follow me relentlessly, but now I think my private FBI surveillance can be for once useful and drive me back to El Paso in one of Mr. Hinkle's new automobiles."

As an afterthought, von Braun offered before he turned to go, "Can you believe they actually think I could be kidnapped by Russia spies? Just look at them," his voice dropped to a whisper, "they use battery lamps when artificial light is not necessary. To guide us tonight, all we need is the light from the moon."

Chapter 43

"It is equal (the manned flight mission to the moon) in importance to that moment in evolution when aquatic life came crawling up on the land."

— Wernher von Braun

The next morning during a sunrise breakfast at the Sombrero Motel's Mexican Cantina, Will Hammond was sitting with Darrell Loan and the two rocket-patch soldiers Jerry Webber and Mike Stark. The two young men were the volunteer photographers from the unofficial observation posts and the bulldozed sand dunes. Both Webber and Stark were obviously pleased to be veterans of the precarious adventure and seen at Hammond's table, and they very pleased to be seated with the nonchalant cowboy from Iowa that *everyone* really did believe redesigned and saved the latest Redstone.

As usual, when the soldiers with the rocket patches were nearby, the little restaurant was a bustle of breakfast activity. Everyone from the previously unexplained civilians in the Chrysler sedans — who were the now exposed FBI — to almost every other technician and engineer that were the proving grounds familiar faces were gathered among the noisy mealtime clatter. Curiously, von Braun, Kurt Debus, and Freddy Shultz were absent.

As the spry, freckled-faced, and unforgettable West Texas waitress dominated the scene and made her rounds with the chuck wagon coffee, Sergeant Will Hammond lit a Lucky Strike, snapped his Zippo lighter closed with a flourish, and settled into his favorite element. He was ready to tell a story.

"That's right boys, "Hammond focused on the younger soldiers and offered a confidential wink. "Why, it wasn't a hundred miles from here, and fifty from the old Proving Grounds, that the flying saucer of 1947 crashed. Of course, everybody knew about it after it happened because the rancher that owned the land came into Roswell and told the sheriff all about it. He brought some pieces of metal into town that day that absolutely . . . were not from this world."

Hammond knew that the two youths were hooked and landed with the latest version of the tale that everyone in the Army *knew was true* and he continued with the voice of a tour guide: "They say those little flying saucer parts looked just like tinfoil but thick. If you rolled the pieces into a ball they would just pop right back out again into their original shape—and there was writing on that silver metal that looked almost like what the old Egyptians used. The Air force of course got a hold of everything after a few days and hauled all the wreckage and whatever else they could find out to Wright-Patterson Air Force Base. They also hauled out, under secret orders, the little gray men they found that were the pilots—but one of them . . . wasn't quite dead."

With the two young men spellbound by Hammond's tale of a doomed expedition from outer space, Bill Hinkle arrived and pulled up a chair.

After von Braun's explanation and the revelation that the latest Jupiter flight had gone off without a flaw—until the unforeseen self-destruct—Bill Hinkle was back to his old confident self. He was suddenly without the heavy yoke of stress and worry, and with the morning light, he seemed cheerful and even tolerant of Hammond's long-winded narratives of the 1940s Wild West. Hinkle was, however, clearly preoccupied with new information; it was bursting beneath his patient smile, but it was also obvious that the new intelligence was not for the sergeant from Huntsville or the rocket-patch soldiers from White Sands.

"Morning Mr. Hinkle," Will Hammond offered brightly, "I was just telling the boys about what happened out there behind the mountains."

"Oh, that's good," Bill said and looked at his watch. "I can't believe the time. I've been on the phone for over an hour — long distance."

In the brief but awkward silence that followed and with Darrell hiding a smile behind his coffee cup, Hammond stubbed out his cigarette and gathered his soldiers who were so very proud to wear the insignia of the U.S. Army Ballistic Missile Agency on the patches on their shoulders.

"Yeah, it's getting late and Captain Freddy Shultz wants us all ready for the road by eight o'clock." Hammond tilted his head to Darrell, and looked at Bill Hinkle. "I guess you two," he said, "won't be riding back to Huntsville today."

"No, no, not today," Bill explained carefully, "we've got some new marching orders."

Little was ever lost on Sergeant Will Hammond, and he knew enough from his experience in the Army when it was time to leave. "Come on boys," he said, rising from the table, "maybe I can talk the Captain into letting you two ride home with me." Hammond then signaled the captivating waitress and motioned for coffee for Bill Hinkle.

"We'll see you when we see you boys," Hammond grinned, "and I've got a feeling it won't be too long."

Before the sergeant and his underlings were out of earshot, Darrell and Bill could hear the beginnings of a new story. "That's right boys, I'll tell you," Hammond began, "strange things can happen out in this desert . . ."

Bill Hinkle shook his head and Darrell smiled over his coffee. After the beautiful waitress with the blue jeans arrived and the man from Chrysler ordered ham and eggs without any peppers, tomatoes, or anything Mexican, he pulled his chair closer.

"Guess what?" Bill asked quietly, but his excitement was unmistakable.

"I don't have a clue," Darrell sipped his refilled and scalding coffee.

"You're flying out of El Paso today at eight o'clock sharp and headed for Audrey and Detroit," Bill confided, and looked around the diner, "but I'm going to visit your old boss Alan Misner in New York."

"What's up in New York?" Darrell asked carefully, but his classic nonchalance faded. "Why does Sperry suddenly want to talk to Chrysler?"

"To tell the truth, I'm not really sure, but Misner sounded determined. By the way," Hinkle added, "you've got a big surprise waiting for Audrey back in Detroit, and don't forget, von Braun promised the White House we'd have a satellite up in orbit in just over a month."

Chapter 44

"The achievements of Apollo were so bold and our subsequent efforts so timid that the energy of those years seems like a youthful dream."

— *Buzz Aldrin*

When Darrell drove the new Chrysler home and parked in the snow-shoveled driveway, Audrey was waiting. She didn't even mind that the new model was a New Yorker. It was a new car, not *too* boxy, and Darrell was home. The New Year it seemed was starting out great and filled with promise. Even the new car looked promising. It was a two-tone—black and gold.

During Darrell's time away at what the other wives called "The New Mexican Proofing Ground," she had taken down the Christmas tree because everyone in the Midwest knew that a Christmas tree up after New Year's Day was bad luck. She had also taken down all of the outside Christmas decorations and she had shoveled the driveway. This was of course because Sally Booth had called and told her she found out from another girl that Audrey and Darrell were getting a new Chrysler for the New Year. Sally also explained cheerfully that new car would arrive every year afterward that Darrell worked for Chrysler and it was very important—as everyone knew—that the driveway must be clear for a brand new car. And with a pot roast, onions, and new potatoes in the oven, and with the whole house smelling ever so wonderful, and with the baby-to-be not making her sick at all in the mornings, she could have not been happier. But she considered briefly — and there was always a "but" in life — but without the Chrysler Wives Club everything might not be so perfect.

When he pulled into the drive, she was watching and waiting because the girl from the dealership in Warren called just after he drove away with the big surprise — and just before the dealership closed — but now that she saw his face, she knew he was worried. He smiled as he always did when he saw her and held his arms outstretched, but she knew he was under a lot of pressure, and when she held him tight, she also knew that in only a few days he had lost weight.

"Okay daddy!" she announced with her best smile. "I know all about the new car and the one to follow this one, and I also know by your look that you could use a nice cocktail and no coffee — and guess what?" she added. "We're having the best pot roast the butcher could cut, and I'm not even going to be upset that after only one day and two nights at our new home in Michigan you have to fly out again, and this time to nasty old New York."

Darrel laughed. "Audrey," he said as he held her at arms length, "what would I ever do without you?"

"Well, you're never going to find out. And by the way, I can tell that you're smoking too much, and drinking too much coffee, and eating who knows what, but now you're home, I love the car, and we're not going to think about Russians or H-bombs or anything bad," she said without taking a breath.

They hugged and kissed, with the quiet snow all around and the new car in the driveway and the wood smoke from the fireplace smelling nice. As Audrey guided her man into their new house and the home she was making, she knew that she would have to put her foot down from time to time and keep him well-fed and safe and somehow away from the too-much-stress that all the wives talked about, as well as all the pressure that came from being what Sally Booth said was "gifted."

"If he wasn't gifted," Sally had explained with a confidential whisper, "he wouldn't have one of the highest security clearances at Chrysler."

Later after dinner, and when the firelight was down to embers, for the first time in days, Darrell slept without the dreams. In New Mexico and Texas his dreams had been of Russians and Communists and their oversized and massive rockets, and the godless fanatics from across the world that wanted to take away everything that he knew was right and good, and his wife, and his family, and his new baby-to-be . . .

Chapter 45

From COLLIER'S magazine circa 1954

A ruthless foe established on a space station could actually subjugate the peoples of the world. Sweeping around the earth in a fixed orbit, like a second moon, this man-made island in the heavens could be used as a platform from which to launch guided missiles. Armed with atomic warheads, radar-controlled projectiles could be aimed at any target on the earth's surface with devastating accuracy.

January 1958

New York City was cold during the second week of the New Year, but not as cold, overcast, and gloomy as Detroit. The early morning TWA flight was delayed because of the January fog in Michigan, but after endless cups of airport coffee, a cheeseburger lunch, and too many cigarettes, the constellation airliner was wheels up over the Motor City and Darrell Loan was flying with a northwest tailwind into New York.

When the plane landed, Bill Hinkle was waiting, and just as in New Mexico, the Chrysler man was in good spirits. After catching up on Audrey, the new home in Warren, and the new car that was the big surprise, and after asking at length about the pretty redhead with the green eyes that was Sally Booth and her equally beautiful little sister Darcy, Darrell and Bill made the drive out over the stone bridges and through the bare trees to arrive under perfectly clear skies in

Long Island. With the late afternoon fading and beginning to cast long shadows, the representatives from Detroit were ushered up and into Sperry headquarters and seated before Alan Misner in his top floor executive office.

Immediately after the airliner touched down in the late afternoon, Hinkle had offered, "Don't worry about the plane being late. I actually think this is better. Alan Misner wants to talk and at the end of the workday there will be fewer distractions."

With the brisk winter day winding down into a frosty twilight, employees of the Sperry Corporation were bundled against the cold, filing out through the main entrances, and moving along the walkways. From Alan Misner's upper office windows, headlights were winking on in the parking lots below.

"Darrell, it's great to see you again," Misner began after a quick welcome from the departing office girls and coffee and ashtrays were in place "And," the Sperry man paused, "to hear about what you have already done with Chrysler."

Darrell smiled briefly and gave Hinkle a sidelong glance. "You were right Alan," he said carefully, "I would have been crazy not to take the job in Detroit. The Chrysler plant at 16-mile is beyond anything I could have expected. Huntsville and White Sands are exciting, and the Germans . . . well . . . they're interesting," Darrell paused. "But I won't forget," he said firmly, "where I started, and how Sperry always treated me right."

The Sperry executive flashed a smile at the compliment but then he was ready for business. "It always goes back to the Germans," Misner nodded as he focused. "And especially von Braun," the man from Sperry suddenly frowned.

"We're at a tricky point here Darrell," Misner went on, "as Mr. Hinkle knows, so just bear with me for a few minutes and we'll see if we can set the stage."

As if coming to a predetermined decision, Misner nodded solemnly, pushed his fingers into a steeple, and leaned back in his chair.

"There was one of von Braun's old associates that really got the ball rolling," the Long Islander explained. "His name was and still *is* Willy Ley—he's living stateside now. This guy was one of the original members of the German Rocketry Society. Of course, this goes way back, before the war, but the Jules Verne idea was simple: To send men out into space, then to the moon, and then to Mars, and to always bring them back again safely. The inspiration was that there would be valuable minerals on the moon for mining and especially on Mars, and that would make the cost and the effort worthwhile. The Germans in those days were really dreamers, but the dream caught on—largely because Willy Ley and von Braun wrote stories about space travel and spaceships to the moon. The articles were filled with wonderful futuristic artwork showing giant rockets with colorful fuel tanks and spider-like spaceships landing on the moon. Those stories and those pictures were published in every language for everyone to see all around the world. That quite simply is why we are where we are today. The whole outer space thing has gone crazy with the world press, the public, and Hollywood, and the idea of a pilot in space is now more exciting than anything—even more thrilling than when the Wright brothers first got off the ground at Kitty Hawk. And, just as when the Wright Brothers first flew, everyone is excited and everyone wants to jump on the bandwagon. The only reason the Russians sent that dog into space was to find out if anything living could survive the G-forces during liftoff. The dog lived until it ran out of air, and you had better believe the Russians right now are working on getting a pilot into space."

Outside the office windows, the Long Island nightfall was complete and Alan Misner on a sudden impulse stood and stretched his back. He then pulled the venetian blinds until there were stars visible out over the bare trees and the Long Island Sound.

"That's the new bandwagon out there," he said and tilted his head to the darkened sky and the stars, "and we need to get on fast—and we can't fail—just because a few Germans want everything done their way. The real problem is," Misner rushed on, "it's not at all about a man in space; it's about showing the world which country can get there first. It's all about which nation has the best engineers and scientists, the best politics and lifestyle, and who can get to the moon."

"It's all about the moon as the bandwagon," Misner reached to the desk and shook a cigarette out from his pack. The Sperry man paused after he lit up and watched the smoke as it curled upward toward the windows.

"All the way to the moon," he continued. "Whoever is leading the race will demonstrate to everyone else that they have the most accurate missiles for delivering atomic war. It's just a contest for the world to see—and a contest of supposedly peaceful technology—but it's really to show the world who has the most talent and the best engineers, and who could win a third world war."

After moment, Darrell glanced to Bill Hinkle who was quietly and patiently smoking and then back to the Sperry executive waiting by the windows. "Alan," Darrell said, "this is really not anything new—a little more out in the open than I ever heard it before—but what do you really want to talk about?"

"We've got a problem and it's a big one," Misner confessed and shook his head, "and this time it's a problem for Chrysler and Sperry all at once. With your background, we decided to bring you in."

"Darrell," Bill Hinkle interrupted, "you don't know anything about what we're going to tell you because, until now, you didn't need to know. But now you do. You need to know *now*, because everything is a little different from what we expected."

"That's right," Alan Misner added, "and what you are about to hear is certainly nothing secret—we've all known about it—anyone with a K&E slide rule and the right numbers can figure it out." Misner frowned suddenly and sat on the edge of his desk. He was waiting for the official denial from Chrysler. "All right Hinkle," he said as he crossed his arms, "you might as well get started."

Bill Hinkle smiled suddenly as if he had been caught with his hand in the cookie jar. He then shifted in his chair uncomfortably, stubbed out his cigarette, and began with the same weak smile, "It's not that bad." He raised his hand as if to wave away the issue. "It's really just a difference of opinion."

"Bullshit Hinkle," Misner flared, "your new Chrysler Jupiter can't get into orbit the way it is and you know it. It needs a second stage to reach escape velocity or that satellite will fall right back down to the ground. Even a small satellite can't get into orbit without that second stage. Without that second stage, the Russians are the only ones riding the bandwagon. If von Braun doesn't make his February deadline and get the Army into orbit, Ike will have a heart attack in the White House and Chrysler just might lose the contract for the Army Ballistic Missile Agency. Isn't that right Mr. Hinkle?"

"Wait a minute," Darrell held up his hand, surprised at Misner's venom and unnatural outburst, "von Braun has already mentioned a second stage to get into orbit. So what's the problem?"

"The problem is," Misner answered in a rush, "is that von Braun wants to use liquid fuel, which has always been his big success story, but liquid fuel has always been tricky and we just can't hope that an untried liquid fuel second stage will work without a failure. Liquid fuels at high altitudes have always been unreliable. That's why the second stage on the Redstone Jupiter, the Army has now decided, will be eleven Sperry-made, Sergeant solid fuel rockets in a circular group, and as a third stage to lift the payload even higher, three more sergeant missiles. The final stage will utilize another Sperry-made Sergeant as the actual housing for the satellite."

Darrell whistled. "Do the Germans know about this new plan?" he asked carefully.

"They do now," Misner looked at his watch, "as of late this afternoon."

Bill Hinkle nodded. "There's been a lot of discussion, and we want you, Darrell, to be the troubleshooter between Sperry and Chrysler. Your job is to make sure the Sperry-made Sergeants deliver their payload into orbit."

"What's the timetable now?" Darrell asked. "How long do we have?"

"That's another tricky part Cowboy," Misner suddenly smiled as he used the nickname he had heard about. "You only have three weeks until it's February, and by the way," he added, "the Germans have decided to replace the alcohol fuel in the Redstone with something they call Hydyne. It's a new and much more volatile, dangerous, and explosive fuel, and no one is really sure where it came from. Isn't that right Mr. Hinkle from Chrysler?"

Chapter 46

"If you are going through Hell, keep going."

— Winston Churchill

"Don't worry too much," Bill Hinkle began in the hospital room, and Darrell decided he had already heard that phrase too many times. "The Navy is scheduled to launch their next Vanguard before von Braun's February timetable is due. If the Vanguard mission works and the Navy makes it into orbit, we'll all get a break in the pressure."

Darrell did not want to think about the pressure; he wanted to think about Florida, the Atlantic Ocean, and warm beaches and palm trees. The pressure from Chrysler and Sperry had put him in the hospital.

After the Misner meeting in New York, the K&E slide rule came out and stayed out for days. Back in Warren, and at the Chrysler plant at 16-mile, the Cowboy from Iowa had been busy. Audrey didn't know there was anything wrong because she didn't know that he was either at Huntsville or somewhere in between, but he had been working around the clock. Misner wanted numbers and calculations. Von Braun and Debus wanted independent proof on paper the trajectory and power of Sperry's Sergeant missiles was what was purported, and unknown technicians at Chrysler wanted backup drawings and more proof and additional emergency contingency plans.

For many days and nights, Darrell had worked under tremendous pressure to finish the endless calculations and to supply all the requirements needed by all the parties involved. After a final third day and night of living on coffee, sandwiches, and cigarettes, the Cowboy decided it was time to go to bed.

All the work was finished and submitted, but as Darrell laid down and tried to sleep, his mind just kept speeding away—numbers, formulas, and sequences, and then a kaleidoscope of technical drawings, over and over and over again.

After hours of trying to relax and fall asleep, sleep would not come. Finally, at 3:00 a.m., Darrell drove himself to the emergency room to see a doctor.

With the doctor unsmiling, listening, and guarded throughout the explanation, the Cowboy was given an injection and admitted into the hospital in Huntsville. He slept for two days.

When Darrell opened his eyes, Bill Hinkle was waiting. He had flown down after hearing the hospital news and his mission was clear. It was time to round up the Cowboy and send him to Florida for the launch of the latest Jupiter Redstone that was now named Juno-1. The Navy was scheduled to launch first, but as everyone was beginning to discover, schedules in rocketry could change fast.

Chapter 47

"Taking a new step . . . is what people fear most."

— Dostoyevski

January 31, 1958 - Cape Canaveral, Florida

The Douglas DC-3 flights from Huntsville to Cocoa Beach became a daily routine during the last two weeks of January, and the last day of the month was no exception. With the new Jupiter-Juno ready and waiting for fueling on the Cape Canaveral launching pad, the key technicians from Chrysler, many of the rocket-patch soldiers from Huntsville, and all the Germans that had recently arrived were staying at a strand of beachside motels on Cocoa Beach. The Germans and the men from Chrysler preferred the Starlight Motel, but all of the engineers loved Cocoa Beach and found the breaking waves of the Atlantic inspiring.

When the sun dipped below the clouds and set inland over the Florida peninsula, the time was just before 6:00 p.m. and just before the liquid oxygen and the new and mysterious Hydyne fuel began pumping into the standing rocket.

"I can't believe the Navy backed out and canceled the scheduled Vanguard," Bill Hinkle said worriedly. "They only gave us five days notice. Only five days to get ready."

"Too late to think about that," Darrell offered as the two men from Chrysler walked though the balmy evening air. The sunset was still aglow in the west, silhouetting the palm trees, but with the last day of January fading, the launch complex at Cape Canaveral was exploding with artificial

169

light. The entire scaffolding and gantry tower was growing brighter with every spotlight focused, and with the preflight activity bustling around the poised and waiting Jupiter, the ebbing day and the encroaching darkness could only be found in the shadows a mile away.

"You're confident about the new fuel?" Bill Hinkle asked, prompted by the sight of the liquid oxygen vapor beginning to vent.

"Oh yes," Darrell's smile was back. The Florida air for him had been like a tonic. "All the static tests in Huntsville went perfectly."

"Did we ever find out about the new fuel — the Hydyne?" Before arriving at the hospital in Huntsville, Hinkle had been in Detroit working franticly.

"What we *did* find out," Darrell lowered his voice as the Chrysler men passed some unfamiliar technicians in the traditional white coveralls, "is that Hydyne is a very complicated hydrogen-based propellant and a formula created for something that we — officially — don't need to know anything about."

"Is this what the Russians are using?"

"All I know is the makeup of this stuff is complicated beyond belief. It delivers much more thrust per pound of fuel, and it was originally designed for something very secret called 'Operation Deep Black Suntan.'"

"That sounds spooky," Hinkle remarked, but then the Jupiter-Juno captured his attention, standing tall beside the gantry tower that was illuminated brighter than daylight. The latest booster from Chrysler and the satellite created by the Jet Propulsion Laboratory in California rose to seventy-five feet. The nosecone of the missile was pointed at the dark Florida sky and a spangle of stars that were visible overhead.

"I'll tell you something else," Darrell was also captivated with the tallest rocket either man had ever seen, "there are so many top secret projects underway that no one could begin to know about all of them."

"Are you talking about The Belgian and the Greek?" Bill Hinkle asked, but he could not take his eyes from the glistening sheet of ice that was forming near the venting liquid oxygen. When the cryogenic oxidizer was pumped aboard at three hundred degrees below zero, the outside of the booster fuselage became covered in a thin coating of ice.

Darrell didn't answer at first but instead focused on the base of the booster and the guidance fins and refractory vanes he could see beneath the engine exhaust cone. After moment he smiled, "The Belgian and the Greek?" Darrell tilted his head with the question.

"You mentioned secrets before I did," Bill suddenly frowned, "And you have the highest clearance anywhere, but everybody in the industry knows there are a lot of secrets going on, and most of them have something to do with aerospace or rocketry."

Hinkle stuffed his hands in his pockets. "I sure hope this rocket works," he said resignedly, "because if it doesn't, we're going to be in trouble."

"I have a good feeling about this one Bill," Darrell was almost close enough to touch a tail fin. "I think about midnight tonight someone is going to make a couple of phone calls and the Russians are going to wake up and have their breakfast ruined. The last thing the Communists want is to hear that we have our own satellite up and running."

"I sure hope you're right," Bill spoke softly as another pair of technicians passed. "Did you know that the JPL in California was ordered to prepare two more Explorer satellites just in case this edition doesn't make orbit?"

"That makes sense, because one way or another, this can't fail. If it does we'll just start over until we get it right. But don't worry too much Bill," Darrell clapped a hand on the Chrysler man's shoulder. "I'll say it again. I've got a good feeling about this one."

"You know what doesn't feel right?" Hinkle suddenly turned and was regarding the two technicians that were also standing apart and looking up at the latest American rocket. "Someone here is waiting to make that report to the Soviets in the same way we saw those slides that came from the Russian launches. Just like von Braun said, this will also be a contest of spies."

Before the Cowboy could answer, a loudspeaker crackled into life and made the announcement: "*Attention! All non-essential personnel please evacuate the fueling area. The upper level winds are down, we are Go for launch of Juno-1, and the countdown clock has started. We are four hours and fifty-eight minutes until blastoff.*"

"Blastoff!" Bill Hinkle exclaimed. "Why in the hell would they call a launch a blastoff? That just sounds like an explosion waiting to happen," Hinkle shook his head. "I can't take another explosion."

"Don't worry too much Bill," Darrell said the same words again and realized he had coined a new phrase. His nonchalance was firmly back in place. "I think there will be a lot of new words coming for the future."

Chapter 48

"There's nothing wrong with change, if it is in the right direction."

— Winston Churchill

With every technician as anxious and poised as the fueled-and-waiting Jupiter, the countdown clock passed through 10:48 Eastern Standard Time and the ignition sequence began on the most advanced Chrysler missile ever built.

When the engine ignited, the fiery blast shook the nearby blockhouse and control room as Juno-1 rapidly began to climb. The Hydyne fuel burned brighter than the previous alcohol mixture, and as the rising Redstone lifted above the gantry tower, the white-hot light streaming from the exhaust cone melted away the Florida night.

As the missile continued to devour hundreds of pounds of fuel and oxidizer with every passing second, the obvious acceleration was incredible. With the pounding sound waves of the controlled explosion shaking the windows of the surrounding suburbs, the bright and rising star climbing over Cocoa Beach illuminated the beaches and cast shadows from the palm trees. The ocean waves were suddenly aglow with the newfound light and the shipping channels had a new beacon that could be seen for miles around.

When the thunder of the rising rocket was past and the burning flare of the booster was well out of sight, the telemetry signals and radio waves of a new age of science was all the information the Germans, the men from Chrysler, and the rocket-patch soldiers had to observe. In the command post and control room of Cape Canaveral, the fretful, worried, and watchful technicians crowded around the electronic terminals that were the radio receivers and transmitters.

As the time ticked through 156 seconds of flight, Ernst Stuhlinger sat in front of his radio transmitter and looked at a stopwatch. His K&E slide rule was out as were the slide rules of most of the engineers. Everyone was calculating the exact moment of the main booster burnout time.

After a glance at his K&E, Darrell slid his slide rule closed and waited for the moment the German at the controls for the second stage would react. The entire room was silent and waiting, and every eye was on a stopwatch or the countdown clock and the secondhand that was sweeping.

After Stuhlinger pressed the signal to ignite the second stage, he nodded to von Braun as the confirmation light deployed on the transmitter panel. Once again, all the slide rules opened in silence as every engineer began their personal calculation for the ignition timing of the untried third stage.

Bill Hinkle was chain smoking and his eyes were everywhere.

At sixty miles above the earth, the Jupiter Redstone booster exhausted its supply of fuel and the rocket engine shut down. Upon reaching engine shutdown, the second and third stages, nestled below the satellite, separated and climbed higher with only the momentum provided by the spent Chrysler booster.

When Stuhlinger's silent radio command reached the highest arch of the flight, the Sergeant missile package burst into life and rapidly increased the speeding payload by thousands of miles an hour. In only seconds, the second stage was depleted and the German pressed the button to fire the third stage into action. After the last button of the series was pressed and the final stage ignited, there was only one sergeant missile propelling Explorer-1 forward and into the realm of space.

When the last of the solid fuel sergeants was spent, America's first spacecraft was traveling over 18,000 miles an hour. The entire launch sequence had taken less than four minutes. After the last command was only a glowing red indicator on the central control panel, the silence in the room was broken and everyone began talking at once—the Germans in unstoppable German and everyone else in a chorus of anxious-to-excited English speculation.

"What do you know about this guy at the controls," Bill Hinkle asked with his hand covering his mouth.

"The same as you Bill," Darrell answered quietly. "He's another of the Boss's boys from the old country."

Suddenly von Braun claimed all attention as he asked for new computations from everyone crowding the control panels. "According to previous calculations," the German slid his slide rule closed, "we should receive a signal from the Goldstone tracking facility at half past midnight. Does anyone have an updated prediction?"

Most of the slide rules were already out and working on the problem of when the satellite would make the first orbital circle around the earth and reappear as a radio signal over Barstow California. The satellite would have to cross above the Atlantic Ocean, then over Africa, Arabia, India and China, and finally over the open Pacific before it would be in range of the big antennas at Goldstone.

Freddy Shultz was translating for Kurt Debus as the room erupted in a new wave of speculation. This time the engineers were confident the Chrysler booster and the upper stages had performed perfectly, and it was only a matter of 90-plus minutes at 18,000 miles an hour before the first circular path was complete and confirmation arrived that the satellite was actually in orbit.

"Twelve thirty sounds right," was a voice near Stuhlinger's control console. Everyone noisily agreed, but as coffee was poured and cigarettes flared from Zippo lighters, the minutes crawled by until the first hour had passed. Well after midnight, the speeding stopwatches slipped silently past the 12:30 deadline and into the minutes beyond.

Von Braun was a statue in the haze of cigarette smoke as everyone waited in a silence that began to become painful. At 12:34 a.m., Kurt Debus began pacing. At 12:36 a.m.,

Darrell slid his slide rule closed. At 12:37 Bill Hinkle looked wrecked.

"If we don't hear from California in five minutes," one of the engineers offered quietly, "we're not going to hear from them at all."

"Quiet please!" When von Braun spoke he didn't sound angry, just determined, "We must not have negative thoughts."

When the synchronized stopwatches swept silently through 12:39, Bill Hinkle looked as if he was going to explode and the air in the hazy room began to feel toxic. No one wanted to be defeated, and after von Braun's warning about soliciting failure, there was not a single individual that was ready to move from the lighted consoles. It was clear that every second was tearing, but there was to be no vocal speculation.

At 12:41a.m., Kurt Debus stopped pacing. He was steadfast when he stopped and he remained silent as he stood watching the console and the technician with headphones that was waiting on the word from California.

Just as the stopwatches swept through 12:42, the technician with the headphones jumped to his feet and began yelling, "Gold has it!" He yelled, "They have it! Goldstone has it! They have it loud and clear! We have a satellite in orbit!"

The room exploded with congratulations and shortly after the backslapping was reaching a peak, the telephones began to ring. One of the calls was from the White House.

Chapter 49

Unbeknownst to the crew in the Cape Canaveral control room, the Jupiter Juno-1 preformed even better than expected and propelled the satellite Explorer-1 into a slightly higher orbit than calculated. The result was a higher circular path around the earth that took slightly longer to complete and reach the Goldstone antennas.

The party really didn't get started until about 2:00 a.m., but when it did, there would be no sleeping for any guests that happened to have a poolside room at the Starlight Motel. The pool deck at the Starlight was the host sight for the party, and when von Braun spoke to the Belgian manager Henri Landwirth and made it very clear that no expense was to be spared for America's first celebration in the race for space, the European hotelier understood his instructions perfectly.

Every possible type of food and beverage was to be made available for the fifty-some-odd technicians and engineers that were responsible for the Juno-1 and Explorer-1 success. It was also made clear there would be no time limit for the extent of the celebration.

Every ninety-six minutes the electronic signal from Explorer-1 would reach the Goldstone antennas, and with every successful orbit complete, a confirmation was relayed to a telephone on the pool deck at the Starlight Motel.

It remains uncertain whether von Braun made the arrangements beforehand for the party or simply called the hotel manager well after midnight and demanded a celebration be prepared at once. Whatever the planning, the party for America's first entry into outer space was the beginning of an era for Cocoa Beach that would last for decades.

When Bill Hinkle and Darrell Loan arrived poolside, the party under the stars was in full swing. As the reports came in with every new orbit completed, cheers erupted over the pool deck and drifted out over the beach and ocean waves. The shrimp cocktail on ice was largely ignored, as were the stacks of every kind of sandwich imaginable, but the bar was busy as everyone continued to celebrate.

When the third orbit was complete and verified, Kurt Debus, with a beer in hand, was thrown fully clothed into the deep end of the swimming pool. With the German technician thrashing about and screaming in German, von Braun and the other Germans began to laugh. After a full minute in the pool and still thrashing wildly at the water, Darrell approached von Braun and asked simply, "What is it that Dr. Debus is trying to tell us?"

"He is saying," von Braun continued to laugh, "that he needs help, because he cannot swim!"

After life rings were tossed to Dr. Kurt Debus and the German technician was rescued from the depths of a Cocoa Beach swimming pool, the party continued until the sun began cresting over the Atlantic Ocean.

The Explorer-1 satellite launched by the Juno-Jupiter-C remained in orbit circling the earth until 1970.

Chapter 50

*"It is a mistake to look too far ahead . . . Only one link of the
chain of destiny can be handled at one time."*

— Winston Churchill

On March 5, 1958, Explorer-2 failed to reach orbit because the final fourth stage refused to ignite, and once again the technicians from Germany, the engineers from Chrysler, and the rocket-patch soldiers from Huntsville were striving forward to perfect the unpredictable as the Navy's pencil-like Vanguard finally reached orbit. When the Navy's mission was complete, the volatile Vanguard had sent a small satellite into space that was about the size of a grapefruit and weighed only three pounds.

On May 15, 1958, Sputnik-3 was launched and gained orbit with an incredible payload weight of just under 2,000 pounds. It was painfully obvious that the Russians were leading in the race for space, and now it was clear to the world that the finish line was on the moon.

On July 29, 1958, President Dwight Eisenhower signed into law the National Aeronautics and Space Act and immediately absorbed the old National Advisory Committee for Aeronautics and all of the related employees nationwide. Also combined into the new agency was every major aeronautical contributor dealing with advanced aircraft, experimental hardware, and most importantly the Army Ballistic Missile Agency. All the engineers and technicians working at the Redstone Arsenal in Huntsville were now part of the newly formed NASA, and the transition was mandatory.

With the formation of NASA, the new agency was unrestricted in any altitude or flight tests and had an annual budget of one hundred million dollars.

With Audrey and the new baby at home in Michigan, Darrell, Bill Hinkle, and Will Hammond stood with hundreds of spectators at the outdoor ceremony in Huntsville as the National Aeronautics and Space Administration was brought into the world on October 1, 1958.

With the arrival of October, Darrell was officially informed that he no longer worked for the Chrysler Corporation; he was now working for the newly launched NASA.

After the initiating ceremony and one of von Braun's most inspiring speeches ever, Darrell crossed with Will Hammond through the Redstone parade grounds. Many of the faces were familiar as the crowd milled away from the ceremonial stage, but now, officially, everything was different. Almost everyone was moving toward the Army Ballistic Missile Headquarters and a building that was soon to have a new sign.

A season of change was definitely in the air as the men and women of the newly formed NASA walked under the clear and cool and endless blue skies of autumn in Alabama. As Darrell and Will Hammond walked together, there was a notable absence as Bill Hinkle was not included.

When the ceremony was finished, Hinkle had offered his hand and congratulations. "I'm officially still with Chrysler, boys, and not part of the new agency. My orders come straight from the top."

Will Hammond scratched his head. "That don't seem right Bill," he offered with his Wild West drawl. "It won't be the same without you."

The Cowboy nodded his agreement but then he smiled. "Don't worry too much Will . . ." Darrell voiced the preamble that had become a new standard. "I've got a feeling our Mr. Bill Hinkle won't exactly be out of the picture."

Hinkle shrugged and stuffed his hands into his pockets. The crowd was beginning to disperse and everyone was filing past. There were furtive glances toward anyone not moving toward the new administration headquarters.

After a sidelong study of a group of passing pedestrians that were a little too close, Hinkle confessed, "There *is* the new Saturn project and a lot of that is under contract with Chrysler. And that will be a launch vehicle that will make the Redstone look like a Fourth-of-July sky rocket," he admitted with a lopsided grin.

"It's going to be the biggest ever," Darrell's comment was not a question.

"That's right boys, and you don't need to worry too much about that. But right now, you two have to get signed up, and I have a plane to catch. And besides," he declared, "the Chrysler Redstone is still the right booster for the new pilot-in-space Mercury project and the Saturn will be the big booster for the future."

Hinkle then offered his classic wink and turned on his heel. In a second, he was away and moving over the grounds, but in the opposite direction from all the others that were headed toward the administration buildings. As he suddenly blended with the crowd, Hinkle reminded Darrell of the two secretive men from Washington that were the agents with the Department of Defense: Mr. Smith and Mr. Jones.

After crossing over the Alabama parade ground and filing past by the military police at the administration building entrance, Darrell was ushered into an office and privately informed by a duty officer that he was now a U.S. government employee and what the command administration called a GS-16.

The officer explained at length that common technicians were now rated and paid as GS-12s. Upper management were labeled and paid as GS-13s, and Wernher von Braun and Kurt Debus were officially GS-18s. The duty officer then explained that all the information was completely confidential, but as a GS-16 with a Crypto security clearance, Darrell was required to know the ranking order of everyone in NASA. It was made very clear that von Braun was still the Boss with many of his Germans holding key positions in the new agency.

When the administration building closed, Will Hammond was standing on the portico beneath the "Army Ballistic Missile Headquarters" and waiting for Darrell. When the Cowboy came down the steps, he nodded to the military police and then grinned when he saw the favorite rocket-patch soldier.

"It's going to feel funny," Hammond began when the two men were out of earshot from the stationary MP's. "I've worn this uniform for years and now they tell me I'm a government employee—but still officially a civilian."

"Yep," Hammond pushed his army cap back to a rakish angle. "It's going to be strange not wearing the uniform. I know it might sound corny," Hammond suddenly looked sad, "but I really like these shoulder patches with the Redstone rockets."

"Don't worry too much," Darrell grinned with the now standard opening. "I was told that at least during the beginning the Army recruits will still wear their uniforms here in Huntsville, but not down at Cape Canaveral in Florida."

"And that's because," Hammond glanced over to the two MPs who were obviously straining to listen, "NASA doesn't want to be considered as having anything to do with the military."

"That's it," Darrell agreed. "We may be using ballistic missiles designed for delivering atomic war, but that's not the focus with our new agency. Our new focus is to explore outer space."

"Then I still get to wear the uniform for a while," Hammond looked hopeful.

"At least up here, but I think we had both better be ready for a shirt and tie or at least some white coveralls with a new kind of rocket patch."

"Hey," Hammond suddenly smiled in the fading October light. "You really think that NASA will have a good-looking patch for the uniform?"

Darrell clapped Will Hammond on the shoulder. "I wouldn't worry too much about that—" he said, "I would count on it. How about a beer and a cheeseburger to celebrate? All this new administration makes me hungry. I'm starving."

Immediately after the formation of NASA, every scientific and industrial entity engaged in the upcoming quest for space began competing for the best method and design for launching a pilot into the place where no one had gone before. Consistent concerns, however, regarding the dependability and safety of the volatile rocket boosters were demanding serious attention from every contractor bidding on the man-in-space project.

Chapter 51

"I have learned to use the word impossible with the greatest caution."

— Wernher von Braun

With the McDonnell Aircraft Corporation chosen as the contractor to build the first American spacecraft ever to carry a pilot, Darrell and Bill Hinkle were ushered into a room full of engineers and test pilots at the Space Task Force complex in Hampton, Virginia. It had only been a few weeks since the inception of NASA, but just as promised, Bill Hinkle was back as a consultant for the Chrysler-made Redstone.

With the blue skies clear, a chill in the air, and the autumn leaves red and gold outside the Langley research center, a classroom-style conference room was called to order. Darrell and Bill sat in the back with the sunshine streaming through the windows. They had a good view. There was always a lot to learn from watching the reactions of anyone invited to a special assembly. The coffee came in Styrofoam cups, but it was much better than most.

There were engineers, test pilots, and technicians from all across the country, and as the fifty-odd-men sat assembled, there was a definite atmosphere of bonhomie and upcoming success. There was also a very real hope for the future. The Russians might be leading in the race for space, but leading and winning, as everyone knew, were two different things. After everyone was seated, a young man with a cautious smile, a bowtie, and an old-fashioned tweed suit appeared at the doorway. He quickly crossed to the center of the room.

"I feel certain," the almost shy and pixyish man began, "that no one here has ever heard of me, because with all that our kind are doing these days, everything is so secret that none of us has ever heard of the other."

Everyone laughed. The young man was instantly likeable, and his voice cultured with an English accent.

"My name is Max Faget, and myself and thirty-four of my colleagues have been hard at work to come up with what we all hope will be a successful endeavor to carry a pilot into space . . ." The bow tie bobbed as the young man smiled mischievously "And of course, to bring him home again safely."

Everyone laughed again and suddenly it was obvious the meeting was going to be a fun assembly. Darrell was pleased to note that Bill Hinkle's tension level seemed relaxed.

"You have all been asked here today," the English accent continued, "to listen to my proposal, to consider what you have heard, and to stop me at any time if there are any questions or points that have been overlooked or remain unclear."

Darrell whispered to Bill Hinkle, "I like this guy."

"Firstly, we must consider that on average only two out of every three of our current rocket boosters are successfully launching without failure or explosion. That, of course, means with the current disaster percentages we cannot possibly ask a man to pilot such a craft without a very good safety precaution."

Bill Hinkle whispered back, "And I don't like him!" Hinkle's tension was suddenly soaring.

"Because of the unfortunate failures — oh by the way," the Englishman suddenly grinned, "I have heard from our various intelligence sources that the Russian booster explosions are even more frequent but they have no free press to report such failures."

"But in any regard," the young man shrugged off the Russian news and began to pace in front of the blackboard. It was suddenly clear the quirky designer was focused, energized, and coming up to speed.

"Because of the danger at liftoff from the booster explosions or failures during early flight, we have designed an escape tower system that can lift our space capsule up and away from a potentially failing booster. With our new design, the space pilot can be away from an exploding booster and at a height of 4,000 feet within only seconds. The escape tower system utilizes a series of small solid fuel rockets. The rockets are attached to a tower atop the capsule, and if there is a problem during liftoff, our pilots can be carried away quickly by the tower rockets and the capsule landed safely with a parachute."

"I must confess," the young man paused and the friendly English voice continued, "the term 'capsule' is just something we've recently come up with, but it seems to be pointing in the right direction. What we've designed is really quite simple because the demands on what our new and special craft must be has eliminated anything but pure function. Let me explain . . ."

The young man with the thin frame and the bowtie paused and smiled as if he were ready to share a secret. He then approached the chalkboard, but he was facing his audience.

"Do we have a representative from Chrysler here?"

Bill Hinkle raised his hand. "Yes I'm here," he said.

"Good," the boyish professor continued, "because we have started where the Chrysler Redstone stops, and I do feel that I must add that the Chrysler-made rocket boosters such as the Redstone are, by far, the most reliable that are produced in America today. The other efforts just don't have the Chrysler reliability."

Darrell nudged Bill Hinkle as the young man began drawing with chalk and sketching on the blackboard. He drew a typical rocket booster but stopped before there was a nosecone or a needle-like-nose at the top of the rocket.

He then began sketching a separate unit above the rocket booster. The image was not unfamiliar as it was triangular in shape, but at the bottom of the conical form he began emphasizing a rounded blunt base that would fit inside the upper collar of the rocket. At the top of his design, he drew a short and stubby round extension that ended the triangular shape, and atop this addition, he drew his scaffold-like escape tower with the small solid fuel rockets at the very peak of his earlier described emergency escape system.

"Here," he said simply as he turned to face the waiting audience, "is the basic design that will house and protect our pilots in space and lift them safely away from a failing or exploding booster."

"The original shape and design idea came from California and an unforgettable character named H. Julian Allen. Harvey, as many of us know him, was given the assignment of solving the critical problem of our ballistic missile nosecones melting away as they reentered the atmosphere. It seems that nuclear weapons, as they reenter the atmosphere at 20 times the speed of sound and 12,000 degrees Fahrenheit, tend to burn up into nothing." The mischievous young man smiled again, "This was when Harvey broke all the rules about how spaceships are supposed to look but reverted again only to function."

"As everyone can imagine, there is no known material that can withstand 12,000 degrees, so, the best way to defeat the overheating to destruction problem is not with a pointed nose but with a blunt and rounded backside."

Max Faget smiled again as he paused for a moment and seemed to consider his own words, but his obvious excitement carried him onward. "Harvey decided that a rounded ball coming back through the atmosphere at Mach-20 would be the best approach, but we've decided a tear-shaped cone with a blunt and rounded backside would be better.

The material on the backside of the cone will be our heat shield and the shape of the heat shield will create a barrier that will protect our pilots from harm. We have also designed a special form fitting 'survival couch' inside the capsule that will hold our pilots in place and enable them to survive the incredible G-force pressure that will be inevitable during the booster launch and the reentry from space.

One of the reasons you have been called to learn of this concept is because we have tested the survival couch with the Navy's centrifuge, and volunteers have determined that survival is possible even up to 20-Gs."

With his lecture apparently complete, the witty young man with the prominent bowtie replaced the chalk, dusted his hands, and stood before the assembled team. He was waiting for a reaction. Before anyone could speak, the English accent continued, "I know that I have popped the bubble on the sleek and sexy X-1 rocket plane destined for space, but what we have now is a design that will work simply and quickly. Are there any questions?"

In the back of the conference room, a young man raised his hand. He was devilishly good looking and had the rugged outdoors look and tan of a movie star. "Did you say that your design and the seat for your design can safely carry a man through 20-Gs?"

"Yes, yes I did" the bowtie bobbed. "That is why we at McDonnell are now under contract to produce 20 of our space capsules that have been designated as Project Mercury. You're a pilot?" Max Faget asked.

"Yes, a Navy test pilot," the young man answered. "I'm Wally Schirra and I'm trying out for the pilot-in-space program."

Chapter 52

"Our two greatest problems are gravity and paperwork. We can lick gravity but sometimes the paperwork is overwhelming."

— Wernher von Braun

High above the Redstone assembly plant in the observation conference room at Chrysler, the two nondescript intelligence agents from Washington were back. As the two men were often in the shadows of every major event or milestone in rocketry, they were referred to as Mr. Smith and Mr. Jones, but their real names were still a mystery.

The spacious observatory was filled with "need-to-know" individuals from Chrysler and NASA, and once again, the blinds to the plant floor were pulled and a slide projector was in place at the center of the room. Bill Hinkle had called and now Darrell and the Chrysler man were seated with the others and ready for anything. The Smith and Jones reports were never good.

"As some of you are aware," Mr. Jones began, and the hubbub of voices from the white coveralls faded away. "The Air Force's planned lunar probe to the moon exploded 77 seconds after the launch of a Thor-Able rocket. This was on August 17. On October 12, the next Thor-Able made a good liftoff but the 'Pioneer' lunar mission disappeared without a trace. The same thing happened on November 8, and now the Thor-Able booster is being reconsidered."

As the agent from the Department of Defense rattled off his report, his partner at the slide projector clicked through slide after slide depicting the Thor-Able rocket in various stages of launch and then episodes of failure.

189

The first slide was a discouraging and violent explosion, just as the Thor-Able was rising alongside the gantry tower. The next image was a photograph taken from a telescope showing the Air Force's launch booster exploding about 70 miles above the earth. The final photos were obviously taken through another searching lens, but there was no evidence of the satellite moon probe, only the darkness of space and a few distant stars.

"Of course," Mr. Jones continued, "anyone that has had a chance to follow the news knows about the troubles with the Air Force and the multiple Thor-Able failures."

As the nondescript Mr. Jones paused to examine the faces around him, Darrell looked to Bill Hinkle and then across the crowded observatory to the other man from Washington who was standing by the projector. There was something wrong. The two men would have not traveled from Washington to deliver an unimportant report. These two were only present to observe critical tests or to deliver bombshell intelligence reports about whatever the Russians were working on. Troubles with an unrelated missile made by another manufacturer were not something drastic enough to warrant all the attention.

The mood in the room reflected frustration. It was clear that most of the Chrysler personnel called to the briefing were irritated about being forced away from their work. There were assignments to be completed, production schedules to be kept, and the news about the Air Force and the Thor-Able mishaps was nothing new. Bill and Darrell, however, were always interested in whatever intelligence the men from Washington had to offer—no matter how ominous—but on this occasion, the information seemed redundant, and that alone sent an alarm signal that was troubling.

"The reason for this latest intelligence briefing," the man from Washington paused, "could be the most disappointing news to date and the end of all rocketry beyond the outer atmosphere."

As the official words from the Department of Defense dropped off into a stinging silence, the self-satisfied man with his indifferent monotone dashed all thoughts of a routine and mundane briefing. Without exception, the entire observatory was now in shock and every engineer galvanized with the unthinkable implication: *Could the space program be over before it really got started?*

Both of the men from Washington appeared pleased with the reaction, and after another lengthy pause for effect, Mr. Jones continued. "What is also compelling, and leading to the same conclusion, is that we have been urged to report that the Russians have also had three repeated failures with their 'Luna' moon probes, and concerns have been mounting that it might not be possible to fully escape the pull of the earth and ever reach the moon. We now believe there is an invisible barrier stopping the rockets."

After the official intelligence statement from Washington and the smug delivery from the agent of the Department of Defense, the observatory exploded into multiple conversations. The denials soon faded into silence as the slide projector reclaimed the room. The man from Washington continued dryly as if the protests from the engineers were only the complaints of spoiled children.

"The Russians have consistently made it further into space than the Air Force's Thor-Able attempts, but at the crucial point of leaving the earth's gravity, the Russian moon probes have all exploded . . . or disappeared completely as did the last two Thor-Able launches."

As the monotone of Mr. Jones' voice droned on, the slide projector was clicking through Russian booster launches similar to what Darrell and Bill had seen before the success of Explorer-1 and the Jupiter-C booster, but clearly, the latest images were updated as some of the smuggled photos were capturing daytime launches.

"Because of the reconsideration of the Thor-Able booster, Chrysler has been given the assignment of launching the next attempt to the moon. The previously successful Jupiter-C model will be launched on December 6, and the mission will be to carry the pioneer moon probe and penetrate the mysterious barrier that seems to be encircling the earth."

After the observatory meeting was over and the Chrysler technicians were still in turmoil about the proposed impossibility of escaping the confines of earth's orbit, Bill Hinkle rubbed his chin and covered his mouth when he spoke. "Well Darrell," the Chrysler man asked, "what do you think?" Hinkle laughed, but his laugh was nervous and filled with tension. "This almost sounds like one of those Saturday afternoon science fiction movies. Is there really someone or something out there? Something that is stopping the moon probes?"

Darrell shrugged as the two friends watched the agents from the Department of Defense. "We're traveling into the unexplored," the Cowboy offered, "But just remember, for a long time, a lot of pilots believed we could never get past the sound barrier and go faster than the speed of sound."

Despite the hostile reactions from the Chrysler engineers, the D.O.D. agents were clearly finished and sorting the slides they were taking from the projector. When the men from Washington snapped their briefcases closed, Darrell walked with Bill Hinkle to the observation windows and began raising the blinds. On the factory floor below, the men and women in the white coveralls were hard at work on the latest version of Redstone.

"I wouldn't worry too much Bill," Darrell spoke the words quietly as he looked over the production lines. "Just like you told me before, those guys are trying to be scary. It's just part of their job."

Chapter 53

"The Soviet Union has become the seacoast of the universe."

— Sergei Korolev

With little Mike and Deb happy and growing up in Michigan, Audrey couldn't believe where the years had gone, but now there was another baby on the way, and Darrell's work at NASA was obviously more in Cape Canaveral than with Chrysler in Detroit. He was away more than he was home, and Audrey knew the work was important, but there was a lot to consider when packing up a family and moving to Florida.

Audrey knew that she would miss all her friends in Michigan, and especially the girls from the Chrysler Wives Club, but Florida always did have a certain appeal . . . especially in winter. Knowing that she and the children would be close to the Cape and all the excitement that was obviously part of the race for space was also captivating.

"Just imagine," Sally Booth had said, "the kind of wives club NASA will have."

Audrey smiled as she packed. It was November again and the gentle snowflakes were falling. But once again, and just as before the children were born, Audrey decided that she would not decorate for Christmas in the old house as a new home was waiting in Titusville. This year, Florida was going to be the place to celebrate the holidays. With a cold winter in Michigan looming, palm trees, beaches, and warm weather breezes were offering an anticipation that was hard to deny.

Ever since the beginning of the work at Chrysler and knowing that Darrell was working on the rockets, Audrey had felt a certain pride for her new family that was somehow connected to a very exciting future. The future was uncertain, however, and with all the talk about Russian H-bombs, atomic war, and fallout shelters, Audrey hoped that Darrell and his co-workers could get ahead of the Communists.

The world was changing fast, and when Darrell came home, there was a definite spring in his step that was really contagious. If America was going to beat the Russians in space, Audrey wanted to be nearby, and she wanted her children to grow up watching the rockets rise and be a part of what everyone was now calling the Space Age. As she picked up an old winter sweater and tossed it on the throwaway pile, Audrey considered she liked the sound of the Space Age much better than the Atomic Age. The Space Age sounded exciting and The Atomic Age very scary.

With packing boxes everywhere and little Mike crawling through one of the open boxes, the telephone rang and Audrey crossed the living room and went into the kitchen.

"Hi Audrey," Darrell spoke over the static on the long distance line. "How's the packing?"

"Good I suppose," Audrey looked across the dining room to where Deb and Mike were playing. "How's the weather?"

"The weathers good, but we've had another setback. We launched the latest Jupiter Juno but she failed to send off her moon probe."

"Hey," Audrey frowned and almost whispered, "can you talk about this on the phone?"

"Don't worry," Darrell answered, "that's old news by now and we just launched yesterday."

"Oh," Audrey gripped the phone tighter, "is this bad?"

"It's just another setback, but there is so much going on, there is no way we can fail. And Audrey, you're really going to like the new house at Titusville. It's next to Cocoa Beach, the Cape, and everything that's going on. The house is a three bedroom, two bath, and perfect for the kids.

"That's nice. I can't wait to see it, and the beach, and the ocean, and to sit in the sun. I really can't wait to take the kids out for a sunbath—just imagine, relaxing in the sun. What are you working on next?"

"We can't talk about that, not on the phone, but I can tell you the plans for the future are out of this world!"

Chapter 54

"Research is what I'm doing . . . when I don't know what I'm doing."

— Wernher von Braun

November 21, 1960 - Cape Canaveral, Florida

With the bright Florida sunshine streaming and the Atlantic waves breaking big on the beach at Cape Canaveral, the very first Mercury spacecraft now sat coupled atop a Chrysler-made Redstone booster. The fueling procedures began at dawn, and as the sun rose out of the ocean, the chill of the late November morning vanished as the checklists and the countdown continued.

The first test flight of the new booster and capsule combination was christened Mercury Redstone One, and everyone that was now a part of NASA was ready, waiting, and anxious as the sweeping hand of the countdown clock continued though the final minutes before launch.

Inside a new blast-proof control room only a football field from the orange gantry tower and the now simmering Mercury Redstone, all the technicians were in place in front of their electronic consoles as the Atlantic surf pounded the nearby beach.

Darrell was in front of the guidance controls as all of the other members of the flight team were in front of their own individual monitors. Von Braun and Kurt Debus were, as usual, pacing from station to station with quick glances at the telemetry information streaming in from the poised-and-waiting rocket.

When the countdown clock was complete at 9:00 a.m., and the ignition sequence began, the true beginning of a manned spaceflight did not perform as intended but instead started an investigation that would be unstoppable until every mystery of the very strange launch attempt was solved. The new Mercury space capsule was not carrying a passenger on the picture-perfect sunny November morning, but if there had been a pilot, the results might have been disastrous.

After only a few seconds of the now familiar roar-of-engine start, the Redstone rocket motor shut down — but the booster carrying the new Mercury spacecraft was already a foot above the launching pad. When the engine abruptly stopped, the Chrysler rocket landed back with a hard thud, but remained precariously upright without exploding. It was the first time anyone had ever seen a rocket fail and fall backward without exploding.

With every technician watching in awe as the Redstone wobbled dangerously with its now wrinkled outer skin, a strange series of events began and revealed for the first time how truly complicated rocketry and spaceflight was going to be. *

When earlier in the predawn light, a technician decided that an umbilical cable was too long so he shorted one of the electric wires leading to the tail of the rocket. This action started a chain of events that was an important lesson that could never be forgotten.

As the Redstone landed hard and rocked back onto the Cape Canaveral pad, it was obvious that the failed booster was damaged, but when Max Faget's escape tower launched automatically, sending the solid fuel rockets and the escape tower soaring into the bright Florida morning, the next sequence of automated events seemed even more amazing.

Three seconds after the escape tower launched, the Mercury space capsule still atop the Redstone booster deployed its drogue parachute and the white silk canopy billowed out in the breeze and then hung dangling beside the still smoldering rocket. After another three seconds, the main parachute deployed with a dramatic pop, and then the emergency parachute three seconds after the main. During the emergency parachute deployment, the entire antenna fairing ejected and then hung suspended by wires from the black McDonnell capsule.

Meanwhile, the escape tower and the solid fuel rockets were exhausted after reaching the designed apex of 4,000 feet and were now falling rapidly back to earth.

Before anyone in the control room and blast-proof bunker could utter a word—because only seconds had passed since the initial launch—a gust of the Florida breeze caught the dangling parachutes and threatened to pull the entire booster and spacecraft over with a full load of fuel, fuel pump propellants, charged batteries, pyrotechnics, and self-destruct charges.

When the launched escape tower crashed onto the beach a few hundred yards away, all of the crew in the blockhouse were still silent and watching in awe as the light breeze played and pulled at the dangling parachutes.

After the escape tower crashed and the final parachute popped, Darrell looked at his watch. The entire event from start to finish had happened in just over one minute.

When it was certain all that could happen automatically had happened, firefighters were called to standby as a waiting game began with the November breeze and the hanging parachutes.

During the next twenty-four hours, technicians waited until the rocket's batteries died down and the liquid oxygen boiled away before they could approach the damaged but still dangerous Redstone.

After deciphering the telemetry readouts that recorded every sequence and relay of the failed launch attempt, the detailed investigation revealed that two electrical cables had parted in the wrong order.

The entire mishap occurred because a single electrical cable was too short and parted 29 milliseconds before electrical grounding was no longer necessary.

When the electrical cable attached to the tail fin parted prematurely ahead of a control cable, the very brief delay set forth a series of events that would be a lesson for all the engineers in the future.

As the booster began liftoff and the cables parted in the wrong sequence, a surge of electrical power rushed through the rocket and triggered a relay that normally signaled engine shutdown at the end of the booster flight. When the engine shutdown order was executed, the smoking Redstone suddenly fell without power back to the pad with a wobble and a crinkling of the outer skin.

When the first relay tripped, the Mercury capsule sensed normal engine shutdown and launched the escape tower with the solid fuel rockets just as programmed. The Mercury capsule, however, was designed not to jettison until the Redstone flight was almost complete or the still-accelerating rocket might hit it.

Because the Mercury capsule contained acceleration sensors, it disabled its own escape from the failed booster because the acceleration devices were sensing a safe constant speed. When the escape tower and its lifting rockets were launched, the parachute deployment was activated and the first drogue parachute deployed. When the tension sensors detected no load on the parachute, the primary and emergency chutes were activated and they too released and popped open.

Before Mercury Redstone One was carefully taken down and disassembled, everyone knew there would be many late nights with burning lights before a human pilot was ready to ride a Redstone.

Chapter 55

"One test result is worth a thousand expert opinions."

— Wernher von Braun

Inside a special housing and training facility at Cape Canaveral, Darrell was watching the monkeys and they were watching him. There were six astro-chimps selected for the pre-human test flights, and after the chimpanzees arrived and their training was underway, everyone wanted to have a look at America's first living creatures that were headed for outer space.

There were serious concerns that the incredible accelerations during liftoff would be too much for a human, or that the micro gravity or weightlessness would affect the decision-making process. Even more concerning was the potentially crippling effect of the multiple G-forces sustained inside the capsule during reentry through the atmosphere. Because the monkeys were trained to push buttons in response to lights on a control panel, the astro-chimps were going to pave the way and test the actual reactions of living beings in space.

Of course, all of the technicians knew that monkeys had ridden rockets before out at White Sands and from the Cape, but there had never been a chimp in a Mercury capsule that had ridden a Redstone all the way up and into space.

The first monkeys were squirrel monkeys, and as everyone now knew, the very first monkey "Gordo" did not survive. His parachute failed to open after reentry but electrodes attached to monitor his breathing and his heart did show that he lived through the flight.

On May 28, 1959, two monkeys were launched together in the same nose-cone-capsule and experienced a suborbital flight as a team. Their names were Mr. Able and Miss Baker. Mr. Able was a rhesus monkey weighing seven pounds while Miss Baker was a South American squirrel monkey weighing only eleven ounces. Their Jupiter flight was a big success and relayed information back to earth that they had ridden the rising booster at speeds of over 10,000 miles per hour, and they had experienced weightlessness for over 9 minutes during their 16-minute flight.

The next major monkey flight was scheduled for January 1961, and because the Russians had already sent satellite probes to the moon and passed beyond the mysterious barrier that seemed to be exploding rockets, confidence was high that an American in space would soon follow the test pilot chimps.

"Did you know that ever since the Russians crashed their 'Luna' probe onto the moon and sent back pictures from the dark side they have been naming mountains and craters after famous Russian cities and Communists?"

"Yeah, I heard about that," Darrell answered the chimp wrangler as the technician finished strapping "Ham the Chimp" into a Mercury capsule survival couch. The training tech then switched on the colored light control panel in front of the soft brown eyes and the furry fingers. Ham's face and hands were exposed, but the rest of the chimp was hidden inside a white fabric suit that could be pressurized. The suit was similar to the suits that jet fighter pilots wore, but the suit that Ham was wearing was called a spacesuit.

"How's he doing with the tests?" asked one of the control room engineers. Most of the men from the blockhouse were now gathered and waiting. They had all heard stories about how clever the monkeys were, and Les Myers, the control room superintendent, had suggested the visit.

"He's doing just fine." The chimp wrangler smiled. "We're ready to start one of the usual training programs and Les wants you guys to watch. He was impressed, but not everybody leaves our training facility with a good impression."

With everyone quiet, focused, and attentive, and now in a huddle around the lighted console and the Mercury survival couch, Ham the chimp sat back and concentrated on his mission. As various lights illuminated on the control panel, Ham clicked the corresponding switches next to the lights. The sequence was fast, complicated, and random. All the engineers were impressed. After about five minutes, the test was complete.

"Does he ever make a mistake?" Darrell asked.

"No, not really, not at this advanced stage. The training is tough, and of course we can make the sequence go faster. We can also simulate flight conditions by placing him inside a boilerplate capsule and making it dark inside—except for the control lights—and we can add vibration." The technician frowned, "If he does make a mistake, he gets a mild electric shock. If he does the sequences perfectly, he gets an apple."

"That does sound tough," another of the engineers spoke up. "Are there any side effects after a bad day of electric shocks and being locked in a dark capsule?"

"You bet," the technician could not stop his smile. "Last month there was a congressman that showed up and found out all about that. He went straight to the top brass and demanded to see one of our chimps."

Before the wrangler could continue, Ham began fidgeting beneath the straps that held him to the G-force survival couch. His furry hands were reaching out and his agitated vocal sounds did not sound happy. He had finished his sequence test and he was waiting for his reward.

"Okay Ham," the wrangler spoke soothingly. The trainer then reached into his lab coat and produced a shiny red apple. When Ham was happily munching away, the story of the surprise visit to NASA by a member of Congress continued.

"I won't mention the congressman's name," the technician suddenly looked serious, "because this is a story that no politician in Washington would ever want to get out."

With all the engineers from the blockhouse baited, Ham was forgotten as he finished his apple and went back to another session of colorized lights and switches. His motivation was clear: after another few minutes of lights with the right reactions, the man with the white coat would make another apple.

"Oh yeah," the chimp wrangler gave a lopsided smile, "that congressman ended up to be a real stinker and in many more ways than one."

"It all started when the phone rang during one of the tests. We had been busy all morning training Enos. Ham weighs 37 pounds, but Enos is a little bigger. He also has a bad temper. He is being trained in a more advanced program and that means a few more electric shocks. When the phone rang, we were asked about showing one of our chimps."

"Ham had just been brought up for his afternoon training, so he was going to be busy. Two of the others were also in training and the other two were with the vet. Enos was the only chimp left that anyone could really visit. But you could just tell," the wrangler shook his head, "that Enos was not in a good mood. He had had a bad morning and the training had been rough."

"I told the brass that this was not a good time and that Enos was not in any mood for a visitor. It didn't seem to matter because, after a couple of minutes, the congressman himself was on the phone and yelling.

"He told me that he didn't care what sort of mood the monkey was in. He said it was ridiculous that I should even consider the monkey's mood. Then he demanded to see what all these tax dollars were being spent on. He warned me that if I didn't show him an astro-chimp that was trained to ride a rocket, he was going to call von Braun."

The trainer shrugged, "I told him to come on down and he could visit Enos in his cage. When the congressman showed up, it was obvious that *he* was in a bad mood. I was trying to get him to settle down by showing him the training facilities, but he kept asking about seeing one of the monkeys. I tried to explain that all of the flight candidates were chimpanzees and not monkeys, but that just made the congressman even more upset. I knew it was a bad idea, about showing him the chimp, because Enos had a bad morning, but I guess Enos didn't care if the visitor was a congressman or not."

"When we walked up to the cage, I knew there was going to be trouble. Enos took one evil look at the man from Washington standing there in his brand new suit and then he cupped his hands behind his backside. Before I could do anything but jump back, Enos passed a bowel movement and threw it in the congressman's face. It was mostly diarrhea.

"After a quiet moment of watching Ham at the colored lights and switches, all of the visiting engineers from the blockhouse cracked up. They could not stop laughing for most of the day."

Chapter 56

"The important thing is to not stop questioning."

— *Albert Einstein*

When Will Hammond arrived from Huntsville, he was unexpected. He had just driven an engine prototype down from Alabama and found Darrell at his new home at the Cape Canaveral blockhouse. Inside the Mercury Redstone control room, Darrell's guidance monitor was in the front row, third from the left.

After the initial surprise of Hammond's arrival, Darrell led a tour of the control room and then outside to the launching facilities at the Cape. The tour continued at a favorite beachside restaurant near the Starlight Motel.

With the Atlantic waves breaking big on Cocoa Beach, the salty air off the ocean was balmy and tangy. Shadows were lengthening over the sand as the winter sun was fading, but there was a big-sky view over the ocean as the breakers crashed and rushed up the beach. As Will Hammond settled in with an ice-cold beer, he began his favorite pastime. He was ready to tell a story.

"Did you hear about the pilots they picked?" Hammond asked after a pretty redhead arrived with a wink and a fresh round of drinks. The rocket-patch soldier watched thoughtfully as the waitress walked away. "They only picked the best seven," he said. "The best seven pilots out of over five hundred."

"I heard something about it," Darrell's classic nonchalance was firmly in place and he had no intention of spoiling the story. Everyone at the Cape had heard about the test pilots that were picked to fly into space, but Will Hammond was much more entertaining when his stories were not ambushed or interrupted.

"Yeah, it's pretty crazy how they ended up with those seven. Everyone's calling them the Mercury Seven," Hammond continued. "They're going to be the pilots that ride those Redstones all the way up to the stars."

Darrell sipped his drink, "Maybe not that far."

"Well at any rate," Hammond was now holding his beer as if it were a rocket booster. He had placed one of the little drink umbrellas into the opening as if it were a Mercury capsule. "Can you really believe they're going to ride on top of one of those monsters when everybody knows that only about two out of three are going up without exploding or flying to pieces?"

Darrell shook his head. "You know that the Redstone is much more dependable than that."

Will Hammond paused to look around the open air deck and then lowered his voice. "I guess it doesn't matter to us now, because we're all working for NASA, but the rumor has it that the Belgian with the Greek project is going to take over."

For a moment, Darrell's nonchalance faded. Then he refocused, one thing at a time. "Tell me what you know about the pilots — the test pilots they picked for Mercury."

Hammond shrugged and settled back to look at the ocean. "Like I said," he continued after a sip of his beer, "they started out with over five-hundred pilots but slowly narrowed it down to a hundred and ten serious applicants. These guys were from all over the states and all branches of the military. But there were strict rules. They all had to be test pilots with at least 1500 hours of flying time. They had to have a jet rating and they had to be less than six feet tall. They also could not weigh more than 180 pounds. And that," Hammond grinned, "was just the beginning."

"But they narrowed it down to the best of the best." Darrell prompted the storyteller. He already knew about the seven specially selected pilots that had been chosen to be the fastest men alive, but Will Hammond's stories never failed to be entertaining or enlightening.

"They narrowed it down alright and it happened fast."

"What do you know about the tests? The tests they put them through."

"Out of the first hundred and ten, only sixty-nine qualified, and those boys were sent to Washington where the medical testing began. The first thing the doctors wanted to find out was how these guys are going to do in space. The monkeys seem to make it all right, but no one really knows if it makes them crazy or not."

Suddenly, Will Hammond arched an eyebrow. "You know," he said, "you really can't talk to a monkey and find out if flying in space will make you go crazy."

Darrell laughed. "Tell me about the tests."

"The sixty-nine that qualified started out with the doctors making them soak their feet in ice water. Then with nearly frozen feet and freezing, they were strapped to metal tilt tables so their hearts and breathing could be checked. This happened while they were upside down. After that they had them running on treadmills until they couldn't take it anymore. Those doctors made them run until they dropped."

"I guess that sent a few home."

"You bet, but the next tests were mental and some of the guys were rejected because their IQ ratings were not genius level or above. Can you imagine that?" Hammond looked incredulous. "First, you have to be a test pilot that can take torture from some mad-scientist doctors, and then you have to be a genius to go to the next level."

"How many did the mental tests send home?" Darrell asked. He was mildly surprised. Will Hammond knew details about the pilot program that were not common knowledge. The operations and control room crews had talked about the strict selection, but Hammond's information was remarkable. Everyone at the Cape knew that these pilots were fearless, but this was the first Darrell had heard about the IQ tests.

"Out of the sixty-nine, thirty-three failed the tests or dropped out on their own." Hammond grinned. "That's when the tricky testing started."

"I've never heard of anything that strict or that tough."

After he spoke, Darrell paused as a bartender passed the table and began lighting Tiki torches. The torches were placed along the deck at the edge of the beach and the twilight setting became Polynesian in the firelight. From inside the restaurant, a boisterous group of young men moved out to the beach deck and were taking a table by the lighted torches. Even with the dim lighting, the crew cut haircuts and the confident manner stood out. This was a group of test pilots that were hoping to qualify for the next selection of Americans that were headed for outer space.

"The tricky testing?" Darrell asked after the newcomers were seated and the bartender was out of earshot.

"Yeah," Hammond leaned forward as a conspirator. "The next tests were sort of embarrassing. There were only eighteen guys left out of the original hundred and ten, and that's when the medical people really went crazy. First, they made the boys drink castor oil — three doses each — then they all had to have enemas."

Darrell frowned and looked thoughtful. "Flying that fast with that much G-force pressure could do almost anything," he paused. "Just going to the bathroom without gravity is going to be complicated. I guess the doctors wanted to make sure these guys have good plumbing."

"That makes sense," Hammond conceded. "But out of the final eighteen — and all of the eighteen passed with flying colors — only seven were chosen. The Mercury Seven were picked because they could work well alone or as team members. They really are . . . the best of the best." Hammond nodded toward the beachside table and the young men with crew cuts. "But NASA is still looking for a few more," Hammond whispered. "Even during all the wars, there's never been a program anything like this."

"Do you know the official term for a space pilot now?" Darrell asked.

"Astronaut," Hammond answered immediately, and then he added, "Do you know where the word came from?"

After Darrell shook his head, the rocket-patch soldier from Huntsville leaned back in his beachside chair and quoted from memory: "The term astronaut comes from the

Greek words for star and sailor. *Astron* and *Nautes*. In English, I guess that makes the new Mercury pilots star sailors."

"Speaking of Greek," it was Darrell's turn to lean forward and speak out of earshot. "Tell me what you know about the Belgian and his Greek Project."

Chapter 57

Atlas in Greek mythology is the God of weightlifting and heavy burdens.

"Congratulations Darrell," Les Myers said. "Do we have a boy or a girl?"

The Cowboy smiled. "A girl," he answered. "We named her Kath. She and Audrey are doing just fine."

"That's great Darrell, just great." Les clapped a hand on the Cowboy's shoulder. "And the other kids, Mike and Deb?"

Darrell nodded. "They're great," he offered, "and excited about a new baby sister."

"That's good, really good. But listen," Les Myers continued quickly, "I know you've had a long night with Audrey at the hospital, but I don't want to wait to tell you the news. The Redstone has been replaced. It's now official. Project Mercury is going forward with the Atlas booster. It's what some of the guys are calling the Greek project."

Les had met Darrell in the hallway to the Mercury control room, and now as the two men walked outside and into the sunshine, the Cowboy's superintendent at NASA began to explain the change in plan that was not good news for Bill Hinkle and the other folks at Chrysler.

"It was really a simple decision," Myers explained, "because it's all about lifting heavy payloads, and about staging reliability."

"But the folks at McDonnell have designed the Mercury capsule to fit on top of the Redstone perfectly." After he spoke, Darrell realized he was not surprised that the rumors about the Greek project were true. "What does von Braun and Debus think about the new booster? The Redstone was their baby."

At first, they were not happy," Les Myers shook his head and it was clear he was reliving a memory. Then he smiled. "But now they've got something new to work on. It's going to be the biggest booster ever made and it's called Project Saturn." Myers was now beaming. "The Russians have a heavy multi-engine booster, and now we're going to have one too—but even bigger than the Russians."

"Okay," Darrell nodded his understanding. He remembered von Braun and the moonlight conversation out at White Sands, and he had heard about the heavy Saturn booster, but the Redstone replacement news was new.

"Tell me all about the Belgian and the Greek." Will Hammond had already related the story, but Darrell was interested to hear the official version from Les Myers. As the two men walked toward the gantry tower and the launching pad that had launched "Ham the chimp" onto the list of successful rocket flights, the early Florida morning was offering the perfect weather for any type of flying. A light sea breeze was coming in from the northeast and endless blue skies reached out over the ocean. As usual at the Cape, there was a scent of salt spray in the air.

"The Belgian has been hidden away ever since 1946," Myers explained as the two men walked. "This was because von Braun has always been in the press. He was the official rocket man. He published his stories and articles about sending men to the moon, and he has been popular ever since. The Belgian is a different story altogether. And it really does prove that in America we don't believe in putting all our eggs in one basket."

"Does this have anything to do with something called Operation Deep Black Suntan?" Darrell's question seemed to catch Myers unaware, and the blockhouse superintendent frowned.

"Darrell," Les Meyers looked around even though there were no bystanders between the two men and the launching pad and gantry tower. "That's something that's only whispered about and nothing to do with us here at the Cape—at least not yet, or not that I know of. But let me tell you about Atlas, or what some of the boys have called the 'Greek project.'"

"As you know," Myers began, "the Redstone booster was perfected by von Braun and the Army and the Vanguard by the Navy, but ever since '46, there has been a Belgian designer working for the Air Force. His name is Karel Bossart. He took the very same rocket research that von Braun came up with during the war and went in a different direction. While the army had von Braun and Debus in Huntsville and Chrysler was manufacturing the Redstone in Detroit, Bossart was working in California to perfect the Air Force's version of a very long-range ballistic missile. Since the Air Force knew that von Braun could produce a dependable rocket for medium-range use, they contracted Convair in San Diego to build Bossart's rocket that could fly 6,000 miles at supersonic speed and deliver an atomic bomb right into the heart of Mother Russia. The plans for the Atlas began in '46 but really got the funding to get off the ground when everyone found out that the Russians were working on their own long-range boosters. This was when everyone realized that traditional bombers—even jets—were too slow when compared to rocketry. But what makes America better," Les Myers suddenly beamed again, "is that the Russians only have one rocket program and we already have three. We have Vanguard, Redstone, and Atlas right now, and a real beauty on the way."

Darrell looked thoughtful. The pencil-like Vanguard really wasn't a consideration. "The latest Redstone just came down from Detroit," he said. "When do we get the first Atlas?"

"The Redstone is for the first manned suborbital flight. That's scheduled for May. Next month we are launching the Ranger-one satellite into orbit with the Atlas. California says the new Atlas booster is already on the way. California says the Atlas is ready for spaceflight and heavy lifting."

Darrell stopped walking and tilted his head toward the vacant gantry tower and the launching pad. "So, the

Redstone was replaced because the Atlas can lift a heavier payload higher and faster. And you mentioned staging reliability?"

"That's right," Les Myers nodded. "Bossart figured out a couple of really good ideas. Only about half of the Jupiter second stages ignite. The Belgian figured out how to light the second stage before the booster ever leaves the pad. Just imagine a pilot light burning and the second stage engine just waiting to power up. The fuel tanks on the Atlas are very lightweight and are held together by the fuel pressure. Over 95 percent of the Atlas weight is fuel. It's the latest ratio and very advanced."

"It sounds dangerous," Darrell leveled is gaze, "and tricky. Do the boys in California have a plan to make the Mercury capsule fit on top of the Atlas?"

Once again, Les Myers clapped a hand on Darrell's shoulder. "They have a special stainless steel collar that some of the guys are worried about, and *that* could be tricky."

Chapter 58

"Every month we do a bold adventure. This is the golden age of space exploration."

— Charles Elachi

April 12, 1961

Far across the globe from Cape Canaveral and the palm trees on Cocoa Beach, an R-7 Russian heavy booster thundered into life and roared over the desert steppes of Kazakhstan. The Russian rocket carried the first human pilot to ever enter outer space. His name was Yuri Gagarin and his face behind the visor of his space helmet was on the cover of almost every newspaper in the world.

With Darrell, Les Myers, and Les Geisman looking over the morning headlines, it was amazing how the American press corps embraced the Communist's success. The newspapers were brutal when it came to the American efforts in the race for space, and it was more than obvious that Russia was the big favorite.

Les Geisman was new to the Cape. He was a fuel and staging specialist and sat in the control room next to Darrell. Almost at once, the Cowboy from Iowa and the big Texan from West Texas had become fast friends. Geisman was huge and big as a bear. He was six foot four, with blues eyes and blond hair. He was German but American born and recently in from Huntsville. When the new fuel and staging specialist arrived, he brought a copy of the *Huntsville Times* that was cargo on the morning plane. With the local Cocoa

Beach paper pushed aside, the three men focused on the discouraging headlines from the home of the rocket-patch soldiers.

Beneath the bold letters reading: **"Man Enters Space"** was the photograph of the Russian pilot and a caption that read: **"Soviet Officer Orbits Globe in Five Ton Ship**."

Another set of bold letters proclaimed: **"So Close and Yet So Far — Sighs Cape Canaveral**," and beneath that: **"US Had Hoped for Own Launch**." At the bottom of the page was a feature article that spilled onto the following pages: **"Reds Win Running Lead in Race to Control Space**."

"Can his ship really be five tons?" Les Geisman asked with his deep Texas drawl. "Hell," he said as he shook his bear-like head, "our little-ole Mercury is not even one ton and we've got to go with Atlas to get that into orbit."

One of the control room technicians called over his shoulder, "There's no way that commie ship weighs five tons — no way! Let's not believe everything we read in a newspaper."

Darrell sipped his coffee and looked beneath the headlines. All of the newspapers he had seen were showing almost the same photograph: The Russian with his fishbowl helmet and headgear either just before the launch or just before he was sealed inside his ship. There were no photos of the launch or of the rocket booster. Five tons was a lot of weight to lift.

"Darrell, you should know . . ." one of the engineers started from across the three rows of consoles. "The new Saturn that von Braun is working on . . . how much do they say it can lift? Isn't that being built at Chrysler where you worked before?"

Before Darrell could answer, Les Myers interrupted, "I can tell you one thing." The superintendent was suddenly addressing everyone in the control room. He frowned and his voice was stern. "We are *absolutely not* going to believe everything we read in the papers and we are *not* going to be intimidated. We have a job to do here and we are *not* going to get sidetracked by outrageous claims on a broadsheet."

Darrell was glad that Les claimed the room. He didn't want to answer and he did not want to think that the Russians could launch five times the weight of a Mercury space capsule. He was also wondering what Mr. Smith and Mr. Jones from the Department of Defense would say about the first man in space being a Russian. There was another troubling thought, and that was the simple fact that the Russians were launching what everyone was calling a ship, while everyone at the Cape was working on launching a capsule. Before NASA began having press conferences, most everyone in America believed that a capsule was something you swallowed when you were sick.

Chapter 59

"Man is the best computer we can put aboard a spacecraft and the only one that can be mass produced with unskilled labor."

— *Wernher Von Braun*

When Darrell went to work, Audrey knew he was going to have a long day. She was at home with the kids and loving Titusville and Cocoa Beach, but today she was worried. She hoped everything would go off without any trouble.

The Florida morning was perfect she decided, and Darrell had commented on the fine weather, but Audrey could not remember her husband ever being so concerned about going to work. Darrell was always Mr. Calm and Nonchalant, but today he was concerned. Today was the first day that an American was going to ride a rocket and fly into space.

The television was flickering black-and-white images in the living room, and Mike and Deb were watching. Kath was in the bassinette and too young to understand that the older kids were sitting on the floor and watching a rocket. The rocket was just standing there and waiting, but the kids seemed spellbound.

Audrey was nervous. She had picked up Darrell's tension as he left for the Cape, and now she wondered if it was safe to let the kids watch television. The fact that millions of other American kids and parents were watching was encouraging, but as Audrey looked up from her ironing board, she wondered how the kids would react if they watched a terrible accident live on T.V. She also wondered what would happen across the rest of the country if this rocket blew up like the others. The other rockets had not carried a man, and if this one blew up with everyone watching . . . and everyone knew there was a man onboard . . . she didn't want to think about it.

On the television, a newscaster was speaking into a microphone. In the background was the waiting Redstone. It was one of the rockets the folks from Chrysler built in Detroit. Off to the side was a group of Florida palm trees. The television reporter was not very close to the simmering rocket with the steam leaking out. The rocket looked dangerous.

The man's voice on the television speaker sounded cheerful: *"Alan Shepard has reported to the control room and the technicians in the blockhouse that everything aboard the Mercury space capsule is A-OK. That's right,"* the reporter repeated, *"Alan Shepard is reporting that the first Mercury mission and Freedom-7 are A-OK."*

A-OK? Audrey wondered. *What in the heck is that supposed to mean?* Audrey was glad she was ironing and had something to do.

The newscaster resumed. *"Of course it has been reported that Astronaut Shepard, who is at this very minute sitting atop the fueled Redstone booster, was bitterly disappointed by the Wisner report. This was the official report to President Kennedy that suggested a manned Redstone mission be delayed. This delay of course resulted in the Russian space pilot being the first in space and astronaut Shepard being forced to wait."*

Audrey knew about that. Darrell had explained that the man who talked to President Kennedy had said if NASA launches a pilot too soon it would be the most expensive funeral a man ever had.

The television voice continued, *"Astronaut Shepard has personally named his Mercury space capsule Freedom-7. Seven because it was the 7th capsule built by the McDonnell contractors and because of the seven test pilots that have been chosen to share in the exploration of outer space. Astronaut Shepard has reported that he did not know why he was selected to go first, but that he was deeply honored."*

As Audrey finished ironing another blouse and looked back to the children mesmerized by the T.V., she considered that the last Redstone had gone up great without a pilot, so maybe this one would go even better. She didn't want to think about what Darrell had said about the other two empty flights that had gone off course and had to be blown up. They had to be blown up or they might have come back down and crashed and killed somebody.

It was amazing, Audrey thought, that the kids could just sit there and wait for the rocket to launch. Normally, they would be outside playing and nothing could keep them inside. There had already been delays and now the newscaster was reporting another holdup. It was clear however that Mike and Deb weren't moving. They were glued to the TV. This was almost like the Saturday afternoon science fiction shows, but this was real.

"Apparently we have just been told that the technicians are now concerned with an above-normal fuel pressure reading and there will be another wait until launch. We are now officially at two minutes and holding . . ."

Earlier, the newscaster had reported and the television cameras had shown thousands of spectators lining the beaches waiting to watch the Redstone launch. According to the news, there were millions more listening on radio and watching on T.V. As Audrey picked up another blouse and checked her iron, she wondered about Darrell at the Cape and if he knew that almost everyone everywhere was watching. She also wondered how many Russian spies were sneaking around Cocoa Beach and working on a plan to make the Redstone blow up.

Chapter 60

*"The excitement really didn't start to build until the trailer —
which was carrying me with a spacesuit and ventilation and all
that sort of stuff — pulled up to the launch pad."*

— *Alan Shepard*

The air in the blockhouse and the control room was filled
with smoke. There were thirty-six electronic monitoring stations and almost all of the thirty-six men seated behind the
flickering screens were smoking. There were ashtrays and
steaming coffee cups stationed on the desks alongside the
monitors, and as Wernher von Braun paced between the
three rows of technicians and their stations, his hands were
in a continuous rhythm of inside his pockets and then outside his pockets. Kurt Debus, as usual, looked older and more
composed, but everyone in the smoke-filled blockhouse was
anxious. There had been a launch delay of almost four hours
because of a faulty electrical relay inside the Redstone.

"We have another problem," announced Gordon Cooper, "other than the real-time trajectory computer." Cooper
was one of the Mercury astronauts and was assigned to the
control room as the capsule communications link to Shepard
atop the Redstone. Everyone could hear the radio transmission between the two men, but NASA decided an astronaut
on a mission would only hear one voice and that would be
the voice of another astronaut. The seven pilots that had un-
dergone the rigors of the Mercury training program were all
fast friends and no one could understand better the danger
and tension of what waited ahead.

"What kind of problem?" von Braun's accent was even more pronounced when he was under pressure. The Boss had stopped pacing and was standing behind Les Myers and Gordo Cooper at the communicator's monitoring station.

"Astronaut Shepard has to go to the bathroom," Cooper voiced the claim as von Braun looked over the young test pilot's shoulder. The youthful astronaut with the crew cut was thoughtfully tapping a pencil on the communications console.

"He has been locked up in there for over four hours, and after a steak and eggs breakfast . . ." After a few quizzical looks, Cooper quickly added, "He has to pee."

"There are no receptacles for such a procedure," the chief medical technician offered without looking up. He was the individual monitoring all of Alan Shepard's bodily functions. His monitor was constantly showing the streaming telemetry of Shepard's heart rate, temperature, and breathing.

"Very well," von Braun decided. "Tell him to wait. Tell him that we are not equipped for such a contingency."

"Dr. von Braun," Cooper shook his head. "He has been waiting for over four hours. He drank coffee and orange juice at breakfast and he has to go. He has to go now."

After a hubbub of conversation erupted and Kurt Debus was consulted in German, it was decided spontaneously that Astronaut Alan Shepard could be given permission to urinate in his spacesuit. The pressure suit after all, was tightly sealed.

With a big grin, Gordo Cooper gave the official go-ahead, and Shepard, lying on his back in the first-manned Mercury capsule, emptied his bladder. In a few moments, the effects of gravity were felt and astronaut Shepard's voice crackled over the control room intercom, "Well, yawl," he reported with a southern accent even though he was from New England, "I guess I'm a wetback now."

There were a few laughs from the men at the monitors but von Braun looked serious.

At two minutes and forty seconds until launch, Les Geisman, the booster control technician, spoke up. He was sitting at his station next to Darrell at the guidance control panel and watching the pressure on the liquid oxygen tank. "Hold," the Texas accent drawled. "Fuel pressure is running a little high. We better hold."

After Shepard, reclining in his spacesuit filled with urine, heard Gordo Cooper relate the dreaded word "Hold," and he listened to the news of another possible delay, he calmly but sternly said into the intercom, "Why don't you boys fix your little problem and light this candle!"

The smoky control room was not quiet. There were multiple conversations, although muted, and the clicking sounds of the electronic equipment, but when everyone heard the stern tone of a Naval Aviator's command coming from atop the Redstone, the clatter from the machines continued but the voices in the blockhouse dropped off and suddenly became sidelong glances.

With Debus and von Braun now standing between Darrell and Les Geisman, the two Germans spoke. After looking at Geisman's screen, von Braun clapped a hand on the big Texan's shoulder and nodded as he announced, "Proceed with the countdown."

Chapter 61

*"I woke up an hour before I was supposed to, and started going
over the mental check list; where do I go from here, what do I do?
I don't remember eating anything at all, just going through the
physical, getting into the suit. We practiced that so much it was
all rote."*

— Alan Shepard

9:34 a.m., May 5, 1961 - Cape Canaveral, Florida

With a clear blue sky overhead and the Florida sunshine
streaming, Mercury Redstone-3 ignited flawlessly and Alan
Shepard reached up to the overhead switch panel and triggered the elapsed time clock. Even though there was only a
slight tremor at booster ignition, Shepard knew his very first
chore was to set the flight chronometer spinning.

After hours of confinement and waiting, the last few seconds of the countdown seemed surreal until Shepard heard
Gordo Cooper announce with the steady voice of a sports-
caster: "We have ignition and liftoff."

"Roger," Shepard reported back to the crew in the blockhouse. "Liftoff and the clock has started."

"Roger, liftoff," Cooper echoed the response.

"Oxygen is Go. Fuel is Go," Shepard reported through
the rapidly rising thunder. "Freedom-7 is still Go."

There was no observation window for the pilot on the
Freedom-7 Mercury capsule but only a submarine-type periscope device for viewing. Just after the vibration of engine
start, but before the sensation of acceleration began, Shepard

looked through his periscope and saw the electrical umbilical cable fall away as the Redstone began to rise alongside the gantry tower.

During the first 45 seconds of flight, the acceleration was smooth, but as the elapsed flight time swept through 60 seconds, an intense vibration began and rapidly increased. At 89 seconds into the mission, Shepard knew that he was in the transonic speed range and passing through the violent turbulence before breaking through the sound barrier.

When the vibration and acceleration shudder began to increase beyond anything forecasted or expected, America's first astronaut could not read any of his instruments because of the intense supersonic tremors. His head and helmet were shaking so violently, Shepard could not determine if he alone inside the capsule was vibrating or if it was the entire Mercury-Redstone combination.

At one minute thirty seconds into the flight, Darrell did a quick calculation. The angle of trajectory and elapsed flight time was all he needed. Shepard was now pressed down into the Mercury survival couch and experiencing 6-Gs. He was touching the edge of space. As the Redstone was rocketing upward and out over the Atlantic Ocean, America's first astronaut was traveling over 5,000 miles an hour. In less than two minutes, all of the flight speed records of the United States had been broken, as no American pilot had ever traveled this fast before.

Shepard's voice was steadfast and determined as he reported over the violence of the shuddering vibration, "Freedom-7 . . . all systems Go."

Everyone in the control room could hear the continuous exploding pulsation over the intercom speakers and the buffeting throb of the rocket engine. It was very clear to every technician in the blockhouse: the man in the McDonnell capsule was fearless.

After 142 seconds into the mission, Shepard heard and felt the Redstone booster shut down. Then he heard the whoosh of Max Faget's escape tower as the tower rockets fired and the escape system was jettisoned. After hearing the escape tower launch, Shepard looked through his periscope for a glimpse of the departing rockets but there was no visual information but the darkness of space. There was, however, a new green light glowing on the control panel indicating the escape tower system had indeed fired without a glitch.

Meanwhile, back in the blockhouse, after the Redstone booster shut down, Les Geisman turned to Darrell and offered, "I'm all done. The Redstone fired perfectly. That was a beautiful rocket," he said with his West Texas drawl.

Darrell nodded but did not comment. His full concentration was on the still accelerating Mercury capsule, the first American in space, and the guidance angles that were now approaching the critical height of the flight trajectory. If the angle for the return to earth and reentry into the atmosphere were too steep, Shepard would be killed with the excessive heat and G-forces. If the angle were too shallow, Freedom-7 would bounce off the outer atmosphere and be lost in space.

At 116 miles above the earth, the first American astronaut looked at his flight instruments. He noted that the temperature outside his Freedom-7 capsule was hovering at around 220 degrees. Inside the spacecraft, the cabin temperature was at 91, but inside his pressurized and very damp spacesuit, the ambient air was a moist but comfortable 75 degrees. Spaceflight was workable.

Before Shepard turned to peer through his periscope, he noticed a single metal washer floating weightless in the cabin. After tucking the free-floating spare-part away for a souvenir, he once again focused on the periscope. There was a gray filter covering the optics that he had switched into place before launch to block the bright Florida sunshine. Just before liftoff, and before the intense vibration began, he raised his arm to switch the gray filter back to clear, but his gloved hand brushed against the "Abort" lever that would trigger the escape tower launching system. After the near mishap of separating the capsule from the booster during the first critical point in the acceleration, a private decision was made that the black-and-white view would be better than no view at all.

Now at his position above the earth, the monochromatic view from the periscope revealed the entire Florida peninsula, the Gulf of Mexico, and a string of islands that were reaching out into the Atlantic. The islands with the noticeable shallow water could only be the Bahamas.

With a burst of static, Shepard reported back to Cooper and the blockhouse, and his words crackled over the intercom speakers. "What a beautiful view!" was the first announcement from the first American in space.

When Freedom-7 separated from the spent Redstone booster, Shepard took manual control and placed the capsule through a series of pitch, yaw, and roll maneuvers. This was before capsule turnaround, which placed the blunt end of the Mercury spacecraft and the heat shield back toward the earth. Freedom-7 had reactive control system rockets to change the attitude and angles of the capsule in the vacuum of space, and when Shepard finished with his maneuvers, he was riding backward down toward the earth.

After the reentry angle appeared on Darrell's monitor, he looked up to the faces that were watching and waiting. Von Braun was staring intently, as was Kurt Debus and Les Geisman. Les Myers was observing intently with Gordo Cooper three seats away. Everyone in the room knew that the next events were even more dangerous than the booster flight.

The Cowboy nodded, "Guidance is good. We're in good shape—right down the middle." Darrell had checked the angles three times. There was no mistake with guidance. There couldn't be if Shepard was going to live through the flight. A mistake with a slide rule could kill a man.

At over 250 miles from the Cape and riding the angle that Darrell was watching on his monitor, Alan Shepard began his descent and once again felt the heavy G-load press him down into the survival couch. The accelerating ride from the Redstone had been thundering and intense, but now Shepard was experiencing a much different sensation with only moderate vibration. The gravity force on the Redstone ride had peaked at 88 seconds into the flight, but now as Freedom-7 was flying backward and racing toward the ocean, the G-force levels rapidly increased along with the outside temperatures. With the blazing friction from the atmosphere turning the capsule into a falling star, and the surrounding fireball melting away the heat shield only inches from Shepard on the survival couch, the G-level forces increased to over 11-Gs and almost twice as much as during the Mercury Redstone ascent.

When it was obvious the G-level force had reached a peak, and the relentless pressure eased, Shepard watched the altimeter slip through 40,000 feet and anxiously waited for the first parachute to deploy and slow the descent. If the descent angle had been too great and the fireball surrounding Freedom-7 had been too hot, the waiting parachutes might be nothing more than melted nylon cords and burned fabric.

During the fiery reentry, communication was not possible with the blockhouse crew. It was a time during the monkey flights when the engineers and flight surgeons thought the worst. An ionization field created by the reentry fireball blocked out all telemetry and vocal communications, and as the static continued to crackle over the intercom, the empty seconds in the control room passed like never before.

When Shepard passed through 30,000 feet, everyone at the Cape once again began to breathe when they heard a fresh burst of static and the voice of America's first astronaut. "Freedom-7," Shepard reported, "passing through 30,000 feet."

After another few seconds, Gordo Cooper was grinning from ear to ear when he heard Shepard's crackling voice relay, "Ten thousand feet and the antenna canister is away and main chute deployed."

As Shepard felt the main parachute partially inflate, he once again peered through the periscope to observe the white silk streaming and then unfurl perfectly into the sixty-three-foot canopy. It was a beautiful sight through the tiny periscope and Alan Shepard realized his prayers had been answered.

Freedom-7 splashed down into the Atlantic Ocean 303 nautical miles down range from Cape Canaveral after reaching a height of 116 miles above the earth. When the capsule splashed into the waves, there was more than a noticeable jolt as the steaming Freedom-7 cooled and tilted toward the water. Inside the blackened and burned capsule, Shepard instantly checked for leaks but found his tortured little ship dry. After a moment of precariously tilting, Freedom-7 stabilized and floated nosecone upright as Shepard made contact with the communications airplane flying overhead.

When the reconnaissance plane Cardfile-23 reported Shepard was alive and well and floating beautifully, the thirty-six men at the blockhouse control monitors erupted with delight.

When Darrell leaned back in his chair and took a sip of cold coffee, he looked around the jubilant control room and realized he had just spent the most intense and longest 15 minutes of his life. The entire flight from launch to splash-down was just under 16 minutes, but an American had flown through the edge of space.

When the intercom crackled minutes later with the announcement that Shepard was safely aboard the recovery ship, and the naval doctors reported that he was disgustingly healthy, it was time to celebrate.

Well before lunchtime, most of the crew from the blockhouse were attempting to drive through the snarled and double-parked cars down to Cocoa Beach. Even with an army of cops blowing whistles and furiously directing traffic, the crowds of eager spectators were crammed into every restaurant and bar or even the slightest suggestion of a parking place. The beaches were thronged with crowds, the ocean side streets packed with milling tourists, but there was an oddity to the new-age crowd as many in the warm Florida sunshine were wearing binoculars hanging from their necks. America had a new pastime and it was suddenly clear that the new spectator sport was watching rockets.

After Darrell parked illegally up and onto a curb—Bill Hinkle style—he and Les Geisman fought their way through newfound patrons of the Starlight Motel. The lobby, the restaurant, and the beachside bar were bursting with the jubilant crowds and the staff was completely overwhelmed. There was no room for anyone that didn't already have a seat or even a place to stand. Above a tribe of carved Hawaiian Tiki gods and a troop of harried bartenders, a television was showing the video of Alan Shepard's Redstone-Mercury launch.

Across the room Geisman saw the flight surgeon trying to signal a bartender. There were easily a hundred people crowding the bar and the doctor didn't have a chance.

"I never imagined anything like this," Geisman spoke above the noisy crowd. "Even with binoculars they could only see the launch for about a minute. I just don't get the attraction. How in the hell are we going to get lunch?"

Darrell shrugged and motioned for the door. When the two friends from the trench were out of the dark Polynesian bar and into the sunshine and the crowds on the beach, both of the engineers knew America was celebrating, and the happy mood was contagious, but it was going to be a long wait for a leisurely lunch of cheeseburgers and French fries on Cocoa Beach.

Chapter 62

"I must admit, maybe I am a piece of history after all."

— Alan Shepard

May 25, 1961 - Cape Canaveral, Florida

When Darrell and Les Geisman arrived at the Cape, the day started out as usual, but certainly did not end that way. The first clue came when Les Myers announced there would be a special televised speech from the president and televisions would be set up in the main conference room for everyone to watch. Les in his ever-friendly manner dropped the hint, "No one in NASA will want to miss *this* speech."

After lunch, everyone from the blockhouse, all the technicians from the launching pads, and almost everyone at the Cape gathered in a large conference center where three televisions were warmed up and tuned in to the same news channel. NASA preferred CBS and an up-and-coming newscaster named Walter Cronkite.

There were folding chairs in neat rows in front of the flickering black-and-white screens, but after all the seats were taken, the latecomers found it was standing room only. Coffee cups and ashtrays were everywhere and almost everyone was smoking. The air was thick with smoke but also filled with anticipation. The room was hazy but it was clear something important was about to happen.

Les Myers crossed in front of the elevated television screens, glanced at his wristwatch, and asked for quiet. "I had a phone call today," he offered after his hands were on his hips. "A phone call that came very early this morning; it was from Washington. I was told to have everyone ready to watch and listen because President Kennedy is about to address a joint session of Congress. I was not told what the speech is about but I was told that it would have a very special meaning for everyone here at the Cape." Les smiled a crooked grin, "Now settle down, shut up, and let's watch the president."

It was obvious that Myers with his last statement was trying to be funny but with the failed "Bay of Pigs" invasion so recently on everyone's mind, there were only a few nervous laughs. There were also worries that the Russians were sending atomic-tipped rockets into Castro's Communist Cuba for a quick sneak attack on the United States. The world had never been closer to all-out atomic war and World War III.

Without further comment, the blockhouse superintendent approached the televisions, turned up the volume controls, and crossed the room to join the rank of technicians standing behind the occupied chairs.

A few moments after the CBS news correspondent announced the President of the United States, a young and energetic John F. Kennedy appeared in a dark suit with a very white shirt and a dark tie. After the introduction and a thundering round of applause, he stood behind a podium. Behind him were Vice President Lyndon Johnson and the Speaker of the House of Representatives. There were microphones everywhere.

The president began, "*Mr. Speaker, Mr. Vice President, my copartners in government, gentlemen and ladies.*" Kennedy paused for a moment and then looked straight into the camera.

"*The constitution imposes on me the obligation to from time to time to give to the congress information on the state of the union. While this has traditionally been interpreted as an annual affair, this tradition has been broken by extraordinary times.*"

The young president paused again for effect and then continued with his strong New England accent, "*These are extraordinary times, and we face an extraordinary challenge. Our strength as well as our convictions have imposed on this nation the role of leader in freedom's cause. No role in history could be more difficult or more important. We stand for freedom."*

As the message to the nation and Congress continued, and as the black-and-white televisions flickered at the Cape, Kennedy proclaimed with a determined and eloquent emphasis that there was a power in the world that was against freedom and the emerging nations that were seeking freedom. He did not name the Soviet Russians as an enemy but his meaning was obvious.

The speech began to take shape for everyone at the Cape when the youngest president ever elected gripped the podium and spoke forcefully:

"It is a contest of will and purpose as well as force and violence – a battle for minds and souls as well as lives and territory. And in that contest, we cannot stand aside."

The president then turned to address domestic issues and recovery from an economic recession. He then delved into the topic of economic and social development abroad, especially with emerging nations.

The next topic was self-defense, the extravagant but necessary cost of self-defense, and the partners in NATO that America counted on worldwide. Then the president turned to civil defense, and how the nation would deal with a surprise nuclear attack. The basis for this part of the message was that no one in their right mind would launch an atomic attack if they knew they would be destroyed equally if such an attack ensued.

"If we have that strength," Kennedy offered, *"Civil defense is not needed to deter an attack. If we should ever lack it, civil defense will not be an adequate substitute."*

The president then explained that a weak civil defense plan was unacceptable as an irrational attack or an accidental attack could cause grave damage to the American public. He then called for funding to triple the amount of spending to add new and improved fallout shelters against the threat of radioactive fallout from atomic bombs.

Next was the hope that a nuclear test ban treaty would be soon signed with the Soviets and ultimately the road for atomic weapon disarmament would begin.

"*Up to now,*" the president said near the end of his disarmament comments, "*The Soviet response has not been what we had hoped.*"

All of the engineers and every one of the technicians at the Cape had been listening and watching intently, but when the president's final segment began, there was no way that anyone at NASA could have been ready for the following statements.

Kennedy now had a sparkle in his eye as he looked into the camera. He was not only addressing a joint session of congress and all of America; he was speaking to the Russians in Moscow. Everyone, everywhere, knew they were watching.

"*Finally, if we are to win the battle that is now going on around the world between freedom and tyranny, the dramatic achievements in space that have occurred in recent weeks . . .*"

Suddenly, everyone knew Kennedy was referring to the Russian pilot in space, and to Alan Shepard's Freedom-7 flight.

"*. . . Should have made clear to us all, as did the Sputnik of 1957, the impact of this adventure on the minds of men everywhere who are attempting to make a determination of which road they should take.*"

There was not a sound in the Cape Canaveral conference room as every eye and every ear was focused on the television screens flickering the same image.

President Kennedy then spoke about advice he had gathered from Lyndon Johnson the vice president and chairman of the National Space Council. He then leveled the field with another verbal bomb for the Russians and a statement that would change world history.

"*Now is the time to take longer strides,*" the New England accent demanded. "*Time for a great new American enterprise — time for this nation to take a clearly leading role in space achievement, which, in many ways, may hold the key to our future on the earth.*"

The young president then acknowledged the obvious Russian head start with their larger rocket engines, and then he praised Alan Shepard's historic Mercury flight.

"We go into space," Kennedy continued, "*because whatever mankind must undertake, free men must fully share. I therefore ask the Congress above and beyond the increases I have earlier requested for space activities, to provide the funds which are needed for the following goals.*"

The president paused once again and leveled his gaze. "*First, I believe that this nation should commit itself to achieving the goal, before this decade is out, of landing a man on the moon and returning him safely to the earth.*"

With the last of Kennedy's words, the entire conference room exploded with multiple conversations. The next few sentences were drowned out with wonder, enthusiasm, and sheer amazement. Finally, Les Myers shouted for quiet as the president outlined his plan.

Kennedy then launched into how much the unprecedented project would cost and how valuable weather and communications satellites would be after they were placed into orbit. He closed a few moments later with the solemn statement:

"*It is hearting to know, as I journey abroad, that our country is united in its commitment to freedom and ready to do its duty.*"

Chapter 63

"He had the rare and precious gift on instilling on his many coworkers his own enthusiasm for hard work and high quality. But he was not only a tough and demanding taskmaster, he was a pathfinder and a problem solver, and he always overflowed with an exuberant joy of life that lighted up many a dark chasms on the road to the stars."

— Ernst Stuhlinger on Wernher von Braun

"Gentlemen, please, some quiet," the Boss had just arrived at the Cape and he was addressing everyone in the conference center. He had just flown in from Huntsville and heard the official news. As usual, Von Braun's navy-blue suit was immaculate and he showed no signs of being travel weary.

"Because of the president's speech, and because our most exciting goal ever has been placed before us, I feel that certain conversations . . . are now necessary."

Darrell was sitting in a folding chair next to Les Gesiman. There were about a hundred NASA personnel present and everyone was anxious to hear what the Boss had to say. Kurt Debus was standing near the entrance next to another German, and it was clear the new technician was translating every word that von Braun would say.

Suddenly the Boss paused and rubbed his hands together. His excitement was tangible and everyone could feel it. He leaned forward and offered simply, "You boys heard what the president had to say," he offered with his accent. "So let's get busy!"

After a moment of silence, the engineers began to whisper and then to talk openly. After another minute, the cavernous room was awash in speculation. Von Braun remained at his station in front of the group but he seemed ready to burst with excitement.

Finally, his hands came up like a conductor before an orchestra tuning their instruments. The engineers and technicians once again fell silent.

"I feel this is too important," the German paused, his hands still held high, "to offer only one opinion about the greatest challenge mankind has ever faced. I would like to open the floor for questions. Your questions I will try to answer, and together we will find the answers that we seek. We are all on the road to new discovery together."

Darrell looked around the conference hall. Everyone was surprised. This was unusual and everyone was unsure of how to react. Les Geisman raised an eyebrow and glanced over the double doors at the conference center entrance.

At the entrance, Kurt Debus was nodding his approval as he listened to the running translation.

In the front row, a junior technician in white coveralls raised his hand. His question was the most obvious, and when he spoke, there were many nods signaling agreement.

"Dr. von Braun," the young man began, "how much do we expect a mission to the moon will weigh? Is it possible to lift such a payload?"

Von Braun nodded quickly, "Yes, yes of course," he said. "But your question has outlined a priority. The president has now allocated a budget to get us to the moon, but what will be important here, is that all of us concentrate on how to reduce the payloads that we send aloft. We must have a weight budget. Our new challenge is to streamline weight and function into an art form. I have no doubt that we can do this together."

After von Braun fielded another hour of questions that led back to weight, payload size, and booster capability, there was another question that was not asked. It was the real question on everyone's mind.

When the meeting broke and von Braun moved through the Cape Canaveral crew, Les Geisman pulled out a fresh pack of cigarettes. It was clear the big Texan was deep in thought as he slowly opened the cellophane, then the foil, and offered Darrell a Chesterfield.

After both men lit up, Geisman tilted his bear-like head and blew out a smoke ring. "Can we do it Darrell?" he looked through the smoke. "Can we really send a man to the moon in less than nine years?"

Darrell looked around the conference hall. The doors were open and the men were streaming out, but the television screens were still showing the newscasters pondering over Kennedys' unexpected goals. Suddenly on the flickering screens was the videotaped recording of Alan Shepard's Mercury Redstone rising from the Cape.

"I know we can do it, if we have enough time," Darrell shook his head and thought of Alan Misner's theory in Long Island. "It's been there all along," he said, "even before the Russians started sending up those moons probes. But the real question is how are we going to do it? It's not going to be as simple as sending up an oversized booster."

Geisman rubbed his chin and loosened his tie. "One thing is certain," he said, "we're not going to the moon with a Redstone or an Atlas pushing a Mercury capsule."

"No, that's for sure," Darrell offered, and then he thought of Bill Hinkle and smiled. "How much have you heard about the new Chrysler, Boeing, and Douglas project? The one they are calling Saturn."

"Von Braun's super-Jupiter booster?"

"That's the one," Darrell agreed and then he looked back to the televisions. The newscasters were still excited. They were showing another Redstone rising beside the Cape's gantry tower. It was another perfect launch.

Across the conference room, von Braun was in conversation with Kurt Debus. Both of the Germans were beaming and their enthusiasm was a lamp that was lighting the room.

Chapter 64

"This morning I drove the back roads to our Mercury launch pad Complex#14. There was a plaque with all our names on it. Now anyone who has his name engraved in marble really has something to worry about."

— Wally Schirra

February 20, 1962

Darrell was in the trench next to Les Geisman. It was called the trench because everyone considered the front row of engineers and technicians in the blockhouse as the first line of defense. If anything were going to go wrong with a rocket launch, the guys in the front row would find out first.

Alan Shepard was patiently waiting a few seats away, but there was a new arrival for the trench that could hardly sit still. The newcomer was from the Convair Corporation in California and his name was Tom O'Malley. Convair built the Atlas. O'Malley really wasn't that new because he had arrived with his first Atlas, but there had been many problems and everyone was tense. The Greek project was unlike the Redstone because the Atlas was proving to be troublesome and unreliable.

Meanwhile, John Glenn was waiting in his Mercury capsule that he had christened Friendship-7. He had been in and out of that capsule many times, and it almost seemed like his flight was never going to get off the ground. He was the only one at the Cape who didn't seem nervous, edgy, or worried. John Glenn had the confidence of Superman.

Darrell sipped his coffee and looked at his monitor. His theoretical guidance path was in place and flickering on his screen and he was once again waiting for engine start. He was curious but also anxious. Despite all the self-assurance that O'Malley was offering, it was clear that the representative from Convair was on pins and needles. There had been so many problems with the Atlas that it was hard to concentrate.

Gus Grissom's second Mercury flight had launched perfectly, but the already proven Redstone had boosted that aloft. The Atlas, however, as everyone was beginning to consider, was fickle and delicate, and wrought with complex problems.

Les Geisman leaned over. "What do you think Darrell? Are we going up today or just sitting here and waiting for something else to leak? Look at O'Malley," the big Texan whispered and enhanced his West Texas drawl, "he looks as nervous as a long-tailed cat in a room full of rocking chairs."

Darrell glanced over to where O'Malley was leaning forward, then behind him to where von Braun was looking over his shoulder. Kurt Debus was standing stiffly behind von Braun with his arms crossed. His body language spoke volumes.

"I think today is the day. Launch day," Darrell offered, but privately he wasn't so sure. There had never been this many problems. The flight had originally been scheduled for January 16, but the big white Atlas with the stainless steel bellyband and the black Mercury capsule cobbled on top had been first stacked together and placed on the launching pad the day after New Year's. That was seven weeks ago.

The first problem had been with the balloon-type Atlas fuel tanks, and that problem had been reoccurring until stormy weather clouded the Atlantic skies and canceled more planned launches. The next launch was scheduled for February 1, but technicians discovered that an earlier fuel leak had saturated an insulation blanket between the fuel and the liquid oxygen containers, and that was another two-week delay. Through opened service panels, the Atlas's fuel tanks were painstakingly repaired and all new insulating material installed.

On February 15, another cold front brought more squalls and windy weather, and the big Atlas was once again forced to wait. The only reliable component of the Atlas-Mercury combination was the astronaut John Glenn. Throughout all the worry and scrutiny, he seemed cheerful and confident that the delays were only part of a destiny waiting to be fulfilled. The man was fearless.

Over the intercom, a new report crackled in that one of the bolts used to secure Glenn into the Mercury hatch had broken and had to be replaced. After the announcement, there were no rumbles of complaint, but everyone in the trench lit a fresh cigarette.

When the sheered bolt had been drilled out and repaired, John Glenn had been bolted into the Friendship-7 capsule for almost four hours. Two hours and seventeen minutes of the four-hour wait had been for technical delays.

With all the beaches crowded with spectators from Titusville to Cocoa Beach and the control monitors flickering through 2:47 p.m., Tom O'Malley from California pressed the launch button and the waiting Atlas fired into life.

After the initial flash of the fiery ignition, and the immediate rush of engulfing smoke and steam, an advancing wall of thunder began pounding through the walls of the blockhouse. When the man from Convair saw that the big Atlas was still in one piece and indeed climbing, he softly offered to everyone in the trench, "May the good Lord ride all the way."

When the bright orange flame exploded from the base of the big white booster, everyone in the control room sat frozen in their seats until the Atlas cleared the gantry tower. There was a stainless steel bellyband rigged around the delicate fuel tanks, another around the top of the rocket to hold the Mercury capsule in place, and with everyone in the blockhouse watching intently and waiting to breathe, there was more concern than with any previous launch. A man was onboard this troublesome rocket and this was not the old reliable Redstone.

Darrell wasn't watching his monitor. He was watching the Atlas rise and begin to tilt toward the ocean. He glanced over to Les Geisman. The big Texan was leaning forward with scrutiny and staring at his fuel pressure, fuel pump, and liquid oxygen readings. The telemetry from Geisman's screen was streaming a plethora of information, but Darrell's job was only beginning. His theoretical guidance and flight path was on the screen, but the actual moment in relation to the theoretical had not begun to register. Even though the first few seconds had seemed like stressful minutes, the Atlas mission for heavy lifting was just beginning.

The Chrysler Redstone had propelled Alan Shepard and Gus Grissom to speeds of over 5,000 miles per hour. It had climbed in under two minutes to over one hundred miles above the earth, but the new Atlas booster was already on the way to hurl John Glenn into an orbital speed of 17,500 miles per hour.

According to the planned flight path illuminated on Darrell's screen, the height of the Mercury-Atlas was to be over 140 nautical miles above the Atlantic Ocean. When the Mercury capsule was above the atmosphere and traveling faster than the earth was turning below, John Glenn would be the first American in orbit. At least, that was the plan.

Just after the Atlas thundered into life, the flight surgeon watching his monitor reported softly to everyone in the trench: "Glenn's heart rate is hammering away, but he is still steady at just over 110 beats per minute." The surgeon shook his head. "My heart would have burst by now—riding on top of that explosion."

When the continuous thunder and the final flames of the rising Atlas climbed out over the Atlantic, the animated control room settled in for the new-age business of space travel. Von Braun and Kurt Debus began pacing in different directions, and Tom O'Malley from Convair finally began to breathe.

Chapter 65

"I don't know what you can say about a day in which you have
seen four beautiful sunsets."

— John Glenn

After O'Malley's softly spoken prayer and the flight surgeon's words about John Glenn's steadfast heart rate, all the technicians were too busy to worry any longer. Everyone was on overdrive.

Glenn was reporting over the amplified intercom, and the pounding sound waves and vibration were very clear over the control room speakers. "Roger," Glenn reported, and his excitement filled the room, "the clock is operating. We are really underway."

"We hear loud and clear," Alan Shepard responded over the rumble. He was the voice of NASA control to Glenn in the Mercury capsule.

"We are programming in roll okay," Glenn's voice was vibrating badly over the intercom, "—it's getting a little bumpy here."

"Roger—standby for twenty seconds." Shepard's tone was rock steady. It was clear; his voice was to be the sound of confidence.

"Roger," Glenn's voice crackled over the speakers.

Shepard was following strict protocol. He began a new countdown for the second chronometer. "3 . . . 2 . . . 1 . . . " he said, "Mark."

"Roger," Glenn's voice was surrounded by thunder, "backup clock has started—fuel at 102-101, oxygen at 78-100, amps at 27."

"Roger, loud and clear," Shepard transmitted, and then he looked over to Darrell at the guidance monitor. Darrell nodded and gave a thumbs up. He then held up a clipboard with the number 69. Shepard responded. "Flight path is good — 69 degrees."

"Roger, checks, okay. Mine was 70 on your mark — have some vibration area coming up here now."

Shepard's response was calming. "Roger," he transmitted soothingly, "reading you loud and clear."

"Roger, coming into high 'Q,'" Glenn crackled over the speakers. Darrell looked to Les Geisman and the big Texan frowned and shook his bear-like head. Both men knew that high 'Q' meant extremely unstable high vibration.

The speakers crackled again. "A little contrail or something just flew by the window out there."

"Roger," Shepard replied but everyone in the trench wondered if something had broken loose and was visible as it shot past the window. The flight surgeon lit a cigarette.

"Fuel 102-101, Oxygen 78-101. Amps 24 — still okay — we're smoothing out some now — getting out of the vibration area."

"Roger," Shepard acknowledged, "you're through max 'Q.' Your flight path is . . ."

Glenn's voice suddenly overrode Shepard's transmission. He was obviously very pleased the vibration was down. "Roger," the speakers crackled, "feels good coming through max 'Q' and smoothing out real fine — cabin pressure coming down by 7.0. Okay — flight very smooth now — sky looking very dark outside — cabin pressure holding at 6.1."

"Roger," Shepard's voice echoed the static-filled response, "cabin pressure holding at 6.1." To everyone in the trench it was obvious that Alan Shepard was the anchor for Glenn to hold on to. There was no reason for Shepard to repeat the cabin pressure reading other than to give Glenn reassurance.

"Roger," Glenn continued, but his transmission was getting weaker. He was already well out over the Atlantic and climbing rapidly away from the Cape. "Have had some oscillations, but they seem to be damping out okay now. Coming up on two minutes and fuel is 101-102. Oxygen is 78-102. The 'G's' are building . . . to 6 . . ."

"Roger," Shepard reported, "reading you loud and clear." Then he glanced over to guidance. Darrell nodded and once again gave a thumbs up. He then held up another indication on the clipboard: P-25.

Shepard quickly nodded, "Flight path looks good. Pitch 25 degrees."

"Roger, back to one-and-a-half 'G's' the escape tower fired. Could not see the tower go, but saw smoke though the window."

"Roger, we confirm staging," Shepard's voice was a rock.

"Still have about one-and-a-half 'G's'." Suddenly Glenn's voice crackled excitedly. "There the tower went right then. Have the tower in sight . . . way out. Could see the tower go—jettison tower indicator is green."

Shepard grinned. "Roger."

"One-and-a-half 'G's'," Glenn's signal was getting weaker.

"Roger Seven. Still reading you loud and clear." As he spoke, Shepard was looking for Darrell's response. "Flight path looks good."

Darrell nodded and connected his thumb and forefinger—OK.

"Flight path still looks good, steering is good."

Glenn crackled, "Roger. Understand everything looks good. 'G's' starting to build again . . . a little bit."

"Roger, Friendship-7. Bermuda has you,"

"Roger Bermuda. Standby . . . This is Friendship-7, Fuel is 103-101, Oxygen is 78-100. All voltages are above 25."

"Roger," the controller's voice from Bermuda sounded strong over the speakers, "reading you loud and clear Seven. Cape is Go. We're standing by for you."

"Roger," Glenn's voice sounded tiny in the expanding distance. "Cape is Go and I am Go. Capsule is in good shape. Fuel is 103-102. Oxygen is 78-100. Cabin pressure is holding steady at 5.8 Amps is 26. All systems are Go."

The elapsed flight time when the Bermuda controller took over the vocal communication was 4 minutes and 25 seconds, and John Glenn in Friendship-7 was traveling over 17,500 miles per hour. He was an American in orbit and moving faster than the earth was turning below.

After Glenn in Friendship Seven passed rapidly over the Bermuda tracking station and continued to fly over Europe, and then the Iron Curtain, there were consistent periods where the orbiting capsule and the astronaut were out of radio contact. Whenever the blockhouse and the crew at the Cape were suddenly back in the communications loop, there was always a burst of fresh telemetry and vocal information from the capsule.

After his second orbit and third sunrise, Glenn was reporting fireflies in outer space.

"That sounds crazy," Les Geisman whispered to Darrell. "Do you think the speed has affected his mind?"

"Just as the sun came up . . ." Glenn was repeating excitedly his claim of seeing insects in space and everyone at the Cape was carefully focused on the words that were being transmitted from beyond the atmosphere.

"There were some brilliantly lighted particles that looked luminous that were swirling around the capsule. I don't have any in sight right now. I did have a couple just a moment ago . . . when I made the transmission over to you."

"We've got more to worry about than fireflies that aren't fireflies," Les Myers was now addressing everyone sitting at their monitors and John Glenn was not privy to the conversation. The control room superintendent shook his head and then tapped his hand on a clipboard that he now held aloft.

"We have a very serious indicator light," he announced gravely. "It seems the landing bag has deployed beneath the heat shield."

All across the bustling control room, lighters flared as the technicians lit up. This news was as bad as it could get. If the air cushion and shock absorber for splashdown had prematurely filled with air, the protective heat shield that would save Glenn from the fireball of reentry was most certainly loose or missing. Without a functioning heat shield, John Glenn was a dead man.

Across the monitors, one of the engineers raised his hand and spoke up. "I am showing the retro pack still firmly in place," he said carefully. "If the retro pack and the three clamps that hold it to the heat shield are still tight, then why don't we just leave the retro rockets on during reentry? I think the landing bag light deploy is just a faulty indicator."

After the structural technician voiced his theory, the control room erupted in multiple conversations. Kurt Debus and von Braun were instantly at the engineer's desk and console along with Les Myers and half a dozen designers from McDonnell.

"Ja Ja," von Braun was nodding as he conferred with Debus and the McDonnell men from St Louis. "Tell Mr. Glenn not to eject the retro pack. The clamps and the attitude rockets will burn up during reentry, but the clamps should hold the heat shield in place."

After three, 90-minute journeys around the earth, and John Glenn knowing for the last half of his flight that he might have a faulty heat shield and that he might be only minutes away from a burning death too horrible to contemplate, Friendship-7 returned from the blackness of space and made a perfect splashdown in the Atlantic Ocean.

A perky and healthy John Glenn and his blackened Mercury capsule were plucked out of the salty brine by a U.S. Navy helicopter and placed on the deck of the USS-Noa that was code-named for the mission "Steelhead."

Chapter 66

On November 22, 1963, John F. Kennedy was assassinated as he rode in an open car in Dallas, Texas.

On November 29, 1963, President Lyndon Johnson issued an executive order. With the approval of the Department of Interior, the new president officially changed the name of the Launch Operations Center in Florida to the John F. Kennedy Space Flight Center, and Cape Canaveral became Cape Kennedy.

Darrell and Les Geisman were in a rush. It was raining at the Cape and all up and down the Florida peninsula, and the tropical storm that was bearing down on the coast was forcing everyone to hurry. There was a new Atlas booster on the pad, an urgent meeting at NASA, and the tropical storm-force winds were picking up.

Darrell was driving a Chrysler New Yorker and Geisman was in the passenger seat. Both men were looking ahead and trying to peer through the late August deluge. From Geisman's side window, a grove of palm trees was bending with the gale-force wind, and in front of the onslaught, the windshield wipers were working overtime. As the Chrysler plowed forward, the rain was coming down in wind-driven sheets.

"The trouble is not about getting there," Les Geisman declared over the rain on the roof. "The trouble is how to land when we do. Project Gemini could make it to the moon, but there would be no way to land."

Darrell was listening, but he was also concentrating on driving. Every raindrop it seemed was big enough to fill a coffee cup, and the windshield wipers could not keep up. There were only a few other cars visible on the coastal highway, and both men were very aware there was an erected Atlas booster exposed to the wind and an urgent briefing at NASA that would determine the course of the mission to the moon. The only other cars on the road could only be traveling to the Cape. No one else would be driving through a tropical storm and gale-force winds.

"Watch out Darrell!" Geisman said with alarm. "You had better slow down. There's nothing we can do about the Atlas on the pad, and if we wreck and end up in a ditch, we sure as hell are going to miss the meeting. We might even miss the Atlas blow over."

The wind suddenly buffeted the big Chrysler and both men could feel the gust off the ocean.

Geisman frowned. "I wonder how the Atlas is doing?"

"By now, the boys have got to have it rigged," Darrell offered, "even if they have to tie it to the gantry tower it's not going anywhere — at least I hope not."

Another impressive draft swept off the ocean and the Cowboy touched the brakes. The car swerved. When the gust was over, he carefully pressed the accelerator.

After a few moments the squall passed and the rain became lighter. "Yes sir," Geisman's West Texas drawl continued, "we've got a real problem on how to land."

"If we make it to this meeting," Darrell paused as he turned up the air conditioner, "we can hear all the options." After wiping at the fogging windshield, the Cowboy glanced at his watch and then shook his head. "I've never missed a meeting," he said, "and I don't want to miss this one."

"Which plan do you like today?" Les Geisman asked. This was Geisman's regular routine and certainly not the first time the big Texan had broached the subject. He had been totally preoccupied with the problem of how to land on the moon. Everyone was, and everyone had an opinion.

"Well, there's one thing for certain," Darrell was looking straight ahead and trying to see through the rain. "There's never been a pilot anywhere that could land on something he couldn't see. All these direct approaches will get us to the moon, but rockets just can't land backwards. They can't see where they're going."

Geisman chuckled. "They sure do it in the movies," he said. "I saw one the other day and this big silver rocket landed perfectly upright on Mars. When the astronauts got out, there were monsters there with three eyes."

Darrell snorted when he laughed. "A rocket might land backward in Hollywood, but not on the moon—at least, not on *our* moon. We have to see where we're landing. There are mountains and craters everywhere and boulders as big as cars—even as big as houses. That terrain is going to be tough. With our guys lying on their backs and looking up, how can they see where they're landing? It's just not practical to land backward."

It was Geisman's turn to laugh, "How about a rearview mirror?"

"Believe it or not," Darrell touched the brakes again, "the idea of landing with mirrors was brought up just the other day."

Suddenly there was another gust and the wind and the rain were back with a vengeance. In the distance the beach was visible, the Atlantic Ocean wild, and the surf on Cocoa Beach pounding. The spray from the surf was as high as the palm trees.

With the windshield wipers slapping out a steady rhythm, Darrell offered, "I like the LOR approach. It just seems like the best idea."

LOR was Lunar Orbital Rendezvous and was currently a hot topic at NASA. Many engineers liked the idea; many did not. After JFK had set the goal for the moon, everyone at NASA had been burning the midnight oil. All the aeronautical builders that wanted the NASA contracts were working even harder. Everyone knew the competition was huge and the rewards fantastic. It was not just the money; it was the prestige of getting the contract to build a moon ship. It was going to be history.

"I don't know Darrell," Geisman shrugged his shoulders and his head brushed the roof of the Chrysler. "LOR calls for two different spacecraft launched at the same time. Two ships at the same time means absolutely precise flying—two ships that will have to dock and connect in space. Precision piloting in space will be like nothing ever done before. LOR means there's going to be a main ship and a separate landing craft. It seems very tricky and more expensive. How could two ships not be more expensive than one?"

"In the long run," Darrell paused, "the LOR could be easier. It could also be quicker to produce and the secret to beating the Russians." As he looked out of his driver's side window, Darrell could see the Banana River. It was a windblown ribbon stretching wide under the rain. Even on the inland waterway, there were big waves and whitecaps cresting on the inland watercourse.

Geisman grunted. "I'm not convinced," he said. "A separate landing machine is something that still has to be designed. That could break the bank when it comes to Kennedy's deadline of landing during this decade."

"Think about this," Darrell turned the Chrysler away from the river and took the approach road to the Cape. He looked as his watch. The time was 8:48 and the NASA meeting was scheduled for 9:00 a.m. "A landing craft opens all kinds of possibilities. A separate landing machine could be designed so that the pilots can look down and see where they're going. The pilot's station could be like in a helicopter with lots of visibility."

"I still say we keep it simple," Geisman was determined. "We keep it simple with one machine and one set of pilots. Two machines means two sets of pilots. It just seems too complicated."

When Darrell parked by the entrance to the administration building, the rain was horizontal and driven by the wind, but in the distance the big Atlas booster was still upright and still firmly gripping it's place on the pad.

"I told you not to worry," Darrell grinned after the two men from the trench were inside and out of the storm. "We still have three minutes before the meeting and the Greek on the pad is still holding up the gantry tower."

When the briefing was over, Les Myers had made the official announcement: Project Apollo was going with LOR or

Lunar Orbital Rendezvous. Myers explained that the innovative idea would be hopefully cheaper and faster to produce.

With all the engineers standing in the administration foyer and looking through the glass for a break in the storm, Les Myers sipped his coffee. "It's not official yet," Myers explained cryptically, "but we have plans for a new building."

Darrell stepped forward. "What kind of building Les?"

"A vertical assembly building," the control room superintendent offered. "A building like no other and a building that will be hurricane rated so we don't have to worry about tropical storms."

Les Geisman looked skeptical. "Are you talking about a building big enough to hold an Atlas and keep it out of the weather?"

"I'm talking about a building big enough to hold a moon ship," Myers offered with a lopsided grin. "And an Atlas," he added wistfully, "will be a baby compared to that."

Chapter 67

"No nation which expects to be the leader of other nations can stay behind in this race for space . . . We choose to go to the moon in this decade and do the other things not because they are easy, but because they are hard."

"But why, some say, the moon? Why choose this as our goal? And they may well ask, why climb the highest mountain? Why 35 years ago, fly the Atlantic? Why does Rice play Texas?"

— John F Kennedy

The First Sunday in August 1964 - Titusville, Florida

When the Sunday papers came to the doorstep it was all about the moon. It was a bright sunny morning but the kids at Audrey and Darrell's house wanted nothing to do with going outside. Deb, Mike, and little Kath were crowded around the breakfast table and the newspapers were everywhere. Even Kath at four years old was enchanted with the newspaper images that now covered the table where the pancakes had been only moments before.

On all of the newspapers, the big headlines were meaningless because the up-close and never-before-seen photos of the moon spoke volumes. Every paper had multiple pages of the new lunar photos because there were so many to choose from.

Success was exciting, and the public relations crew at NASA had fun with the press release. There were over 4,000 photographs that had been sent by radio waves from out-

er space. Ranger-7 had six television cameras, and the unmanned lunar probe that had launched from the Cape four
days earlier had been totally successful. The robot investiga-
tor that was about the size of a small sailboat had traveled
across space to send back hundreds of high resolution images before crashing into the lunar surface and making a new
crater. The photos were like nothing ever seen before and
revealed for the first time how truly alien the lunar world
was going to be. Planet Earth was captivated, and now there
could be no doubt: human explorers traveling to the moon
would be the greatest adventure anyone could ever imagine. The photos from Ranger-7 made the Sunday headlines
in almost every newspaper in the free world, and everyone
stopped to take notice.

There were gray craters everywhere with steep rims and
ancient crumbling walls. There were even smaller craters
that were obviously newer excavations inside the larger and
older bastions that looked as old and weathered as the pyra-
mids. There were gray oceans of powder that appeared as
vast lakes trapped between jagged mountain ranges, and
there were giant fissures and deep crevices that seemed to
hold hidden secrets. Consistent among all the lunar photos
was the essence of an untouched and timeless mystery and a
lost world concealed with the shadows of desolation.

It didn't take much imagination for the children to picture themselves tracking through the powder on the gray lunar surface and exploring the mountains on another world.
Darrell smiled as the kids rifled him with questions. Watching the rocket launches and seeing a missile disappear was
one thing, but the newspapers were now showing what
happened when the math and science worked and uncovered the unknown.

Before the newspaper pictures, Audrey had never imagined there could be so many variations of the color gray.
There was, of course, light gray and dark gray, but there
were mixes of the black and white monotone that tended
to run toward snow cream and black ice. There were other
sharp cracks of white that looked like the surface had been
shattered by something tremendous. Audrey, like everyone

else, had always looked at the moon, but until today she had no idea what she was really looking at. It was almost disturbing that something so openly beautiful at night could have so much harsh and hidden mystery, as revealed by one of Darrell's rockets. There was no doubt about it: the 1960's were the future.

Audrey knew about the Ranger missions with the television cameras, and she had seen the launch. It was hard not to know about the rockets when you lived near the Cape. Whenever there was a launch, the ground began to tremble, the rumbling thunder began, and all the windows in everyone's house shook until the booster was well away and up over the Atlantic. Darrell had said that even after only a few minutes, the rocket was up in space, but it just didn't seem possible that something you could walk out the door and watch could be in the blackness of outer space in only minutes.

"Dad," Mike began as he poured over the newspaper photos on the kitchen table, "tell us about the moon and the rocket that took the pictures."

Darrell smiled behind his coffee cup. He knew that the children were interested in all the happenings at the Cape, but this was a new experience. He suddenly realized that almost everyone in America was having the same breakfast table adventure.

"There's a lot know," Darrell said, "and a lot more to find out."

Audrey couldn't stop her smile as she poured more coffee. That, she thought, was a very good answer.

"What are all the 'X's' for?" Deb asked. On all of the lunar images there were black "X's" or "T's" marked on the Ranger photos. Once again Darrell understood that all across America other kids were probably asking their parents the same questions.

"Those are just reference points Deb," Darrell sipped his coffee, "so that the folks that start mapping the moon can know where one picture starts and another one ends. Mapping the moon is going to be very important."

"What about all the rough spots?" Mike said suddenly. "It looks like the moon might not have too many good places to land."

Darrell was pleased. Mike got right to the problem. "The moon is going to be tricky," he offered, "there's no doubt about it, but the closer we get the more we'll know."

"Are we really going to get there soon?" Audrey could not stop her question. She knew that Ranger-7 had been the seventh robot probe sent to photograph the moon, but she also knew that the first six Ranger missions had all failed or had been lost in space. One out of seven was not very good odds.

Darrell paused after Audrey's query, but for only a second, and then he smiled. "You bet."

When the Cowboy turned back to the newspapers and more questions from the kids, Audrey frowned. She didn't want to think about a group of astronauts in a little submarine-type capsule and an accident that could leave them gasping for air and lost in space.

Chapter 68

"If the mission didn't succeed . . . we would have held up the
whole program."

— *Wally Schirra*

October 25, 1965 - Cape Kennedy

Darrell was in the trench and looking at his slide rule.
The old K&E was battered, scratched, and dented, but the
slide worked perfectly and the numbers were easily visible.

Les Geisman leaned back in his chair and watched as
Darrell rechecked a calculation. Both men were on standby and studying their monitors as they waited for the most
important Gemini mission to date. Today, there were two
rockets scheduled for launch and they were only 6,000 feet
apart. There had been another technical delay and the tension in the blockhouse was as heavy as the cigarette smoke
that crowded the room.

"Darrell," Geisman began as he stretched his back, "why
don't you get a new K&E? That thing is worn out."

"It's my good luck charm," the Cowboy declared. "As
long as I can still read the numbers, I'm keeping this one."

It was true, Darrell considered, the old K&E *was* good
luck. It had traveled from Sperry in New York and Chrysler
in Detroit to Huntsville in Alabama, and it had successfully
checked the guidance programs on all of the monkey and
Mercury flights launched from the Cape.

When the Mercury program was completed, Gordo Cooper had climbed with the new and improved Atlas booster and rode in his Mercury capsule for 22 orbits. His mission aboard the Faith-7 lasted over 33 hours, and he had been the last American to travel into space alone. Darrell and Les Geisman had sat side by side in the trench for the whole flight—33 hours. They had dozed, but they remained at their stations—everyone did.

As the blockhouse settled in for the latest delay, the teletype chattered in the background, the weather reports droned over the overhead speakers, and as usual, Kurt Debus and von Braun paced along the rows of flickering monitors. There were three rows of stations and 36 monitors. There wasn't a mission launched from the Cape when the two Germans weren't pacing the rows and quietly speaking German.

What was fun for Darrell was that Les Geisman spoke the language, and whenever the Boss and his chief technician would pass, Geisman would give a whispered translation—but only if it was something interesting. Les Geisman was a Texan—born in West Texas, but his parents were both German and born in Germany. The Geismans always spoke German in their West Texas home.

After his spectacular finish for the Mercury program, Gordo Cooper had ridden the new Titan booster that flawlessly propelled him and Pete Conrad up and into orbit in the new two-pilot Gemini capsule. Together, and continuously circling the earth, the two astronauts had proven that an eight day mission to the moon was possible. It would take Apollo three days to reach the moon and three days to come back. Two days had been set aside for men to walk on another planet and explore the lunar surface.

Everyone knew that the new Apollo program would have much more room onboard and would carry three astronauts, but as the latest Gemini and Titan booster were preparing for flight, it was amazing to realize there was an upcoming December mission that was scheduled for fourteen days in space. Two astronauts were going to spend two full weeks orbiting the earth in a Gemini spacecraft that had a passenger compartment about as big as the front seat of a sports car.

According to the original seven astronauts, the compact Mercury capsule was something that was worn like a coat rather than something that was actually ridden in. For better accommodation and much longer flights, the two-passenger Gemini was a little more than double the size of Mercury and had survival couches for two pilots. The latest spacecraft also had a service module behind the command capsule and storage for a new kind of food supply that was squeezed out of tubes.

The Gemini spacecraft, however, as everyone at the Cape now knew, was a beautiful design and a wonderful little ship. All the Astronauts loved to fly her. The ease of maneuverability and response to the flight controls was incredible. The McDonnell-made Gemini flew like a jet fighter in space.

Les Myers claimed the room as his voice crackled over the intercom. "We are Go and T-minus one hour until Atlas-Agena liftoff. The countdown is on."

"Busy day," Les Geisman said as he leaned back in his chair and looked at his second monitor. Les had two monitors and so did Darrell. Almost everyone in the trench today had two screens. It was the beginning of a new era.

The first screen was showing all the streaming telemetry for the unmanned Atlas-Agena. The Agena was the robot rocket that astronauts Wally Schirra and Tom Stafford were going to dock with in orbit. The Atlas-Agena was to go up 90 minutes before Gemini-6, and then the big Titan booster would launch Schirra and Stafford aloft just in time to meet the Agena after it completed one full circle above the earth. The goal for the mission was to prove that one spacecraft could rendezvous and dock with another. This was an essential maneuver if Americans were going to utilize Lunar Orbital Rendezvous and make it to the moon. If Gemini-6 couldn't dock with the unmanned Agena, Lunar Orbital Rendezvous was not going to work.

The second monitoring screen was showing all the information about the two-stage Titan booster with the black Gemini capsule on top. The two-man Gemini was different from the Mercury capsule as it did not have the familiar es-cape tower and emergency rockets atop the nosecone. If there was booster trouble aboard a Gemini mission, the two pilots were to eject with an explosive charge that would propel them at rocket-fast speed away from a troubled Titan. The ejection seats—typical of that in a jet fighter—looked good on paper, but on one occasion when the ejection system was tested, the heads of the crash test dummies had been badly crushed. The astronauts were all fearless, but no one wanted to think about the ejection seats in the Gemini.

Les Myers was back on the intercom. "Gemini-6 is load-ed," he reported over the speakers. "Wally and Tom are all strapped in."

Chapter 69

"It's a real sobering feeling to be up in space and realize that one's safety factor was determined by the lowest bidder on a government contract."

Alan Shepard

Les Geisman was tapping a pencil. "The Titan is a real beauty," Geisman declared. His sharpened pencil then pointed to the screen that was filled with fuel and pressure readings coming in from the Gemini-Titan combination. The big Texan then nodded toward the flight director's main monitor and the televised image of Gemini-6 waiting on the pad in the beachside Florida sunshine.

Beneath the black Gemini capsule was a white cone-shaped service module attached to the silver Titan booster. Contained in the service module collar that was part of the Gemini spacecraft were fuel cells that produced oxygen, water, and electricity for extended flights through space. If America was going to the moon, the fuel cell technology had to be proven in the upcoming Gemini missions.

"Just look at this Darrell," Geisman was pointing to the data on his favorite screen. "All the readings on the Titan are perfect. Very little fluctuation with anything, and even when there is a problem, it's predicable."

"Yes sir . . ." Geisman's Texas drawl proclaimed. "These Martin-made Titans sure are dependable . . . *and* predictable."

"Don't forget the Greek," Darrell pointed his K&E. "We have an Atlas back on the pad." The Cowboy was indicating the flight director's monitor as the television crews switched back to the other waiting rocket.

All during the preflight preparations, and when a loaded booster finally reached the end of the countdown clock and the ignition sequence became a launch, everyone in the trench looked to the nearest flight director's television.

With the two rockets ready for launch and the gantry towers only six thousand feet apart, the television crews were having the biggest day ever. On the constantly changing screen was the now standard Atlas booster waiting and boiling-off liquid oxygen vapor as a flock of seabirds flew by the standing white rocket and the orange gantry tower. Fixed atop the Atlas was the slender Agena robot rocket and docking vehicle. In the distance, and in the background, the Atlantic swells were rolling and breaking on the beach.

"Don't worry too much," Geisman offered. Darrell had taught the big Texan the classic phrase he had learned from Bill Hinkle. "According to Convair," the Texas accent continued, "Atlas is now new and improved, but the fuel pressure readings are still all over the place compared to the Titan. I'm telling you, Darrell, the Titan is predictable. It's the best booster we've ever had."

"The Titan has to be. It's the front line of defense for all of America," Darrell said as he checked a calculation and glanced over to the countdown clock and secondhand that was sweeping. "I wonder how many people watching this on T.V. realize they are watching a ballistic missile? A booster that was never designed to lift men into space, but a workhorse rocket built to deliver atomic bombs."

The Cowboy nodded to the flight director's television. The newscasters were once again focused on the manned Gemini-Titan combination that was stacked, loaded, and simmering on the pad. Inscribed in bold black letters vertically on the silver booster were two words: "United States."

Les Giesman leaned back in his chair and cracked the knuckles on his bearlike hands, "The Russians know what they were designed for," he winked, "because the Communists *are* predictable and they certainly are watching this today. They're watching everything—especially today."

"Our pilots sure aren't predictable." Darrell glanced over to the grainy image of the black capsule atop the standing silver fuselage. "There's no telling what they'll do, or what's going on inside that little ship. These boys sure are entertaining."

"That's for sure," Geisman raised an eyebrow and looked down the trench. Everyone was waiting and watching the monitors. The cigarette, cigar, and pipe smoke was thick. Moving among the three rows of flickering screens were two attractive young women dressed in the light blue flight attendant uniforms from Pan American Airlines. The "Flight Attendants" were pouring fresh coffee and serving sandwiches and snacks. Pan American had the catering contract for NASA, and the two young ladies swept though the control room on regular intervals. Whenever there was a countdown delay, the Pan Am girls made an extra run.

"I wonder if there are any corned beef sandwiches smuggled in on this flight." Darrell looked over to Astronaut Alan Bean who was the vocal and intercom link to Wally Schirra and Tom Stafford in the stacked and loaded Gemini-6.

One of the flight attendants was smiling as she poured coffee for astronaut Bean. Her brunette hair was tied up in a bun, and her smile was engaging. She was as beautiful as a magazine cover and her uniform was perfect. With her hairstyle and uniform, she looked very futuristic.

The astronaut in the trench today looked like everyone else. He was clean cut and wearing a white short-sleeve shirt and a tie. If Bean knew about any pranks, it wasn't showing. He had the perfect poker face when he wasn't cracking an ever-ready smile and joking with the girls from Pan Am.

Everyone knew that the Astronauts were becoming more and more unpredictable in their pranks and practical jokes. Officially, NASA was not impressed and scolded the pilots for their boisterous behavior, but, unofficially, there were winks and smiles whenever there were unplanned activities that made *almost* everyone in the control room laugh.

Gus Grissom and John Young were the first to fly in the two-passenger Gemini, and after they easily achieved the first of their three orbits, Young produced a corned beef sandwich from Wolfie's on Cocoa Beach. Against all the rules, Wally Schirra — one of the original seven astronauts — had smuggled in the sandwich from the popular deli and had helped Young hide the delicatessen in the front of his spacesuit.

Once into orbit, Grissom took a couple of tentative bites somewhere over China, but soon there were crumbs floating like tiny asteroids around the tight little cabin. After another two orbits with more floating morsels forming their own solar system, Grissom maneuvered Gemini-3 into the standard backward approach for reentry but overshot the landing site by 60 miles. After splashdown when the two pilots were floating on the waves and waiting for the recovery ship to arrive, they were joking over the radio and calling their downed spacecraft: "The Unsinkable Molly Brown."

Naming the downed Gemini capsule "The Molly Brown" was Grissom's way of making fun of himself. For whatever reason, his "Liberty Bell" Mercury capsule had been destined for the Atlantic seafloor when explosive bolts detonated prematurely and allowed seawater to flood in and sink Grissom's little ship. The Mercury program's second astronaut was noticeably embarrassed as he was hauled dripping from the Atlantic swells after his abandoned capsule sank beneath the waves.

Alan Shepard, after the first Mercury mission, had been given a tickertape parade in New York City. After the parade, America's first astronaut and his wife were presented to President Kennedy and the First Lady in the White House. Gus Grissom was the second astronaut in space, but after his capsule was lost, his only reward was a quick celebratory ceremony and an official inquiry on the loss of the spacecraft. NASA was tough.

After Gemini-3 and the nicknamed Molly Brown mission with the corned beef sandwich, Gemini-4 launched effortlessly to an entirely new audience and a whole new mission.

Gemini-4 was referred to as "Little Eva" because the flight was having an extracurricular activity. The "Little Eva" action was a real showstopper and something the Russians had tried only weeks before. The Russians might have gone first, but the Communists didn't understand on how to capitalize on a sensational situation.

When Ed White opened his hatch in the vacuum of space and left the capsule for a heart racing space walk, Planet Earth was captivated. After propelling himself to the end of a tether with a T-shaped handheld rocket system, White, in his spacesuit, with the helmet and reflective sun visor, became the first American to touch the untouchable and experience spaceflight without a spaceship. Ed White, with his arms outstretched, traveled almost halfway around the world before returning to the safety of his Gemini capsule. The Russian cosmonaut Alexey Leonov spent 12 minutes and 9 seconds outside his spacecraft, but Ed White spent 36 breathtaking minutes soaring above the earth.

White's partner and Gemini pilot James McDivitt photographed America's first walk in space as 12 European countries televised the mission via Early Bird satellite. The Soviet coverage for their space walk had been grainy and limited, whereas America filmed and televised its opening space walk in living color.

With the sapphire haze of the atmosphere an earthbound entity, and the deep blue oceans and the tan and sandy continents passing below, it was suddenly obvious to the world that no country or ideology could have secrets from anyone in space. There were no boundaries or Iron Curtains visible from beyond the atmosphere.

Everyone in the trench had watched the flight director's color television, and everyone agreed: Gemini-4 was a real showstopper. The Communists in Russia hated Gemini-4.

Gordo Cooper and Pete Conrad rode shoulder to shoulder into orbit on Gemini-5, and after a few hiccups with onboard fuel cells and an accidental dump of freeze-dried shrimp floating around in the capsule cabin, the two astronauts remained in orbit for a week and a day, and proved that an eight-day mission to the moon was possible.

The two boisterous pilots designed a mission patch that was not loved by NASA, but with a wink and a nod, an old covered wagon from frontier days was a new rocket patch approved for a new era. A rickety Wild West wagon was at the center of a circular crest with an emblazoned caption that offered: "Gemini-5 — Cooper — Conrad — Eight Days or Bust."*

With the red secondhand of the countdown clock sweeping through the final minutes before the Atlas-Agena launch, everyone in the trench settled in to concentrate. Never before had there been two launches scheduled 90 minutes apart, and there had never been two rockets with so much explosive power only 6,000 feet away from one another. NASA was ramping up, and the space race was getting faster.

*NASA managers objected to the "Eight Days or Bust" rocket patch, feeling it placed too much emphasis on the mission length and not the in-flight experiments. Because of concerns that the public might see the mission as a failure if it did not complete the full eight days in orbit, a nylon cloth was sown over the rocket patch until the end of the mission.

Chapter 70

"The earth is too small a basket for mankind to keep all its eggs in."

— *Robert Heinlein*

October 25, 1965 - Cape Kennedy

With thousands of spectators once again lining the beaches of Titusville and Cocoa Beach, the Atlas liftoff was perfect. The Agena docking vessel was in place above the Atlas booster, and as onlookers applauded the thundering liftoff, they were anxious to see the manned Gemini capsule with the Titan booster follow. This was to be the first attempt at docking one spacecraft with another in preparation for the Apollo moon missions that were already into the planning stages. If the Lunar Orbiting Rendezvous was going to become a reality, it would require a lunar landing vehicle to make multiple rendezvous and docking maneuvers with an Apollo mother ship.

"We are two minutes thirty five seconds into flight," Les Myers confirmed.

"Flight status?" the Flight director's voice sounded harsh over the intercom. Today, the flight director was Christopher Columbus Kraft, and after the words were spoken, Les Myers anxiously looked over to his troops in the trench.

"Booster is Go Flight," Geisman was all business.

Darrell followed with the same precision. "Guido is Go Flight."

Down the line the voices continued.

"Electrical is Go Flight."

"PA is Go Flight." This was the polished and official voice of Gemini Control. His was the authorized voice of NASA that reported to all the television and radio stations. His name was Jack King. Jack was a former reporter and journalist now working at NASA, and his post as the Public Affairs Officer was well considered. The brass at NASA figured it takes one to handle one, and Jack was about to get busy.

After Tom O'Malley pushed the ignition button and the Atlas-Agena began to climb above a thunderous exhaust flame, Tom Stafford and Wally Schirra were well into their checklists and waiting for liftoff in Gemini-6.

Launch pad leader Guenter Wendt and his White Room crew dressed in their standard dust-free white coveralls, surgical caps, and white shoe covers had strapped the two pilots into their survival couches and closed and locked the spacecraft hatches fifteen minutes before the Atlas liftoff.

Onboard Gemini-6, the preflight checklists were right down the middle, the redline values perfect, and the powerful Titan booster beneath the two astronauts fully fueled, waiting, and ready for orbital pursuit.

The Atlas liftoff and flight into the supersonic was beautiful and inspiring for the spectators watching from Cocoa Beach, but after six minutes into the unmanned mission, there was no further telemetry offered from the Agena-robot rocket. Tom O'Malley's Atlas booster had apparently performed without a flaw, but the unmanned Agena was in trouble. All across the Atlantic Ocean radio waves searched, but all that was detected was a few articles of falling debris found on a distant radar screen. The Agena docking ship was lost, and Gemini-6 would have to be postponed. After separating from the Atlas booster, the Agena had mysteriously exploded.

When it became official that the Gemini-6 mission was scrubbed and the Atlas-Agena missing, Public Affairs Officer Jack King was in a very bright spotlight until his explanations satisfied the relentless press corps. It was a very long day.

Chapter 71

" . . . The United States was not built by those who waited and rested and wished to look behind them. This country was conquered by those who moved forward . . . and so will space."

— John F. Kennedy

Guenter Wendt, the Gemini pad leader, spoke to everyone who would listen. "Oh, man. You are crazy," he offered when he saw the hanging Titan booster and the advanced schedule for launching Gemini-6 only eight days after Gemini-7. There had never before been such a fast turnaround on the same launching pad. The new schedule called for an Air Force helicopter to airlift and deliver the second stage of an already assembled Martin-made Titan booster.

After the mysterious Agena explosion and the postponement of Gemini-6, everyone at NASA had been very busy. An extremely aggressive idea had cropped up, and the overwhelming fact that the new plan was real was now hovering in the sky and ready to land at the Cape. No one had ever seen a giant sky-crane helicopter lift a rocket booster, but lately at NASA there had been a lot of new firsts.

The plan began with two McDonnell contractors and two astronauts. Wally Schirra and Tom Stafford were still strapped in and waiting for launch atop Gemini-6 when Frank Borman and Jim Lovell, who were the scheduled pilots for Gemini-7, began to open the hatch on a new and very aggressive plan.

Immediately after the Atlas-Agena failure and the press clamoring for answers that Jack King could not possibly have—at least not at the spur of the moment—the two astronauts for Gemini-7 met after leaving the outside viewing stands. Both of the NASA men were headed for the blockhouse and the Gemini control room to see what happened. This was only minutes after the loudspeaker announcement that the Agena had been lost. When the two pilots entered the blockhouse, the control room was packed. Everyone was speaking and there were a hundred conversations.

Darrell and Les Geisman were still at their stations in the trench with Kurt Debus and von Braun standing behind them. Geisman and the Germans were speaking German at a breakneck pace, but Darrell could understand all too well the rapid-fire questions about the lost Atlas-Agena and the sidelong glances Kurt Debus was giving Tom O'Malley at the far side of the monitors.

Meanwhile, over the crowd of technicians, the haze of cigarette smoke, and the open chorus of speculation, two of the McDonnell contractors responsible for the Gemini spacecraft were standing in a corner of the control room having their own conversation. As Borman and Lovell were filing past, dressed like tourists just off the golf course, they heard the words that would change the pace of the American space program forever.

"Why couldn't we launch another Gemini as a target docking ship instead of an Agena?" The McDonnell man that asked the question was spacecraft Chief Walter Burke. He was asking his assistant John Yardley but before Yardley could answer, the two passing astronauts were suddenly spellbound with the new idea.*

As the big Air Force helicopter lowered the Titian booster and the Cape Kennedy ground crew ran to secure the hanging rigging ropes, Les Geisman shaded his eyes from the sun. "She's a real beauty Darrell," Geisman offered to his best friend from the trench. "Those folks from Martin do an incredible job."

Darrell knew that Les was not exaggerating. On every Martin-manufactured Titan booster there were endless tests that absolutely guaranteed quality. As the booster crew dressed in the deep blue Martin coveralls gathered at the base of the airlifted Titan, everyone from inside the control room was standing to watch.

"They found over three hundred inspection points that failed in the first Titan booster," Darrell offered as he too shaded his eyes and watched the hovering helicopter and the Martin crew at work. There were at least fifty men in the blue coveralls labeled "Martin," and they were scrambling to make ready Gemini-6.

"Let's hope there aren't any points that fail during this fast turnaround launch. This is not our style Darrell," Geisman added carefully. "When you rush you make mistakes — and mistakes in our business can equal disaster."

There had never been a plan to have two manned Gemini missions up at the same time. The original plan was to determine if docking or maneuvering next to another spacecraft was even possible, but that was the purpose for the unmanned Agena.

Four astronauts and two Gemini missions up at the same time had never even once been considered. There were tracking ships and ground stations all over the globe that would have to be modified.

The struggle to convince NASA, the flight directors, and all the ground support worldwide that two Gemini spacecraft maneuvering in orbit together would be a big step forward was not an easy achievement.

After 51 days of challenges, debate, and round-the-clock planning, the most complex mission ever was ready for a launching schedule and a Cape Kennedy plan that would make international headlines for Christmas.

Chapter 72

*"I had the ambition to not only go farther than man had gone
before, but to go as far as it was possible to go."*

— Captain Cook, 1776

"Of course, no one knows what the surface of the moon
will be like. We can speculate and we can guess, but we
won't know for certain until we get there. And that could be
trouble." The Cowboy, Les Myers, and Les Geisman were at
a briefing hosted by the Grumman Corporation. They had
flown into New York City and were now at the Grumman
manufacturing plant in Long Island.

The speaker with the Long Island accent was a large-
framed man in a dove gray suit. He was one of the engineers
in charge of the lunar landing vehicle. The Grumman designers had been awarded the contract to build and design
the excursion vehicle that was going to land on the moon,
but there were obstacles that the Grumman people were
concerned about. The concerns were startling, and the obstacles were considerations that no one had thought about
before.

Darrell moved over as Les Geisman stretched out in his
little chair. Geisman was uncomfortable. He was far too big
to sit in the small theater-type seating.

On the stage below, the Grumman official was speaking
and using a pointer. He was directing attention to a map
of the moon that was projected onto an oversized viewing
screen.

The theater-style lecture hall was dark with the only lighting the full moon on the screen. From the stage, the Grumman man's voice sounded as strange as his otherworldly comments.

"As we can all see from the photos from Ranger-7, the lunar surface is a very complex and treacherous challenge. What is even more concerning is what might lie beneath. We have no idea what will be found under the surface of the moon. The moon could be hollow and there could be giant caves."

"Oh Brother," Les Geisman whispered and then shifted again in his chair. "This is beginning to sound like something out of the Saturday afternoon movies or an excuse for delays and more money."

Les Meyers leaned forward and gave Geisman the look: *Be quiet or we'll talk about this later.*

"No one can be sure if the lunar surface can support a landing vehicle." The Grumman engineer was tapping his pointer. He was indicating the darker valleys that were a deep gray color between the rugged mountains and the pockmarked craters. The craters were everywhere. There were even smaller crater rims crumbling inside the larger ones.

"Many scientists are quite frankly worried that the sandy areas in between the mountain ranges are oceans of dry quicksand. These areas might appear solid, but if these educated concerns prove valid, any attempt at landing might be a fruitless effort and like trying to drive a car on a lake. The landing vehicle simply might disappear into the lunar sand and be lost. The Sea of Tranquility could be nothing more than cosmic dust. We don't want to lose any men on the moon."

The man in the gray suit paused with his pointer. He was illuminated by the projected image of the circular lunar surface. The moon looked huge to the audience of engineers and the man from Grumman small and insignificant beside the goal set by John F. Kennedy.

"Are there any questions about the lunar surface?"

The observation theater at Grumman was spacious, but there were many empty seats. It was clear the Grumman representative was well practiced and had preformed the same lecture many times. It was also clear that he was learning of new concerns with every experience. If anyone had valid worries over what might happen when men in a spacecraft were approaching the lunar surface, it would be the technicians and engineers from NASA.

Les Geisman shifted again in his chair and raised his hand. He then realized the man on the stage could not see anyone beyond the front row of the dimly lit theater.

"Ah . . . I have a question," Geisman's strong Texas drawl broke over the sparsely filled seating. Les, when he chose, spoke perfect High German. He had degrees in both chemical and aeronautical engineering and he had been a pilot since he was sixteen. In addition to his formal education, his street-smart common sense was uncanny. The big bearlike man also had the ability to look at a complex problem and see at once into the heart of the issue, but, mostly for fun, whenever he was in aerospace meetings with strangers, he always enhanced his southern accent. He had used this technique many times to ambush unsuspecting colleagues from Northern provinces such as Harvard, Princeton, MIT, and especially anyone from New York.

"Yes, do we have a question?" the Grumman man's Long Island accent suddenly sounded impatient. When someone spoke with a southern accent the New Yorkers usually bristled.

"I was just wondering," Les expounded, "why do we think the moon is dead and the earth has life? Do we think there could be something poisonous up there? I mean, after all, why wouldn't the moon have plants and life too if there was not some kind of toxic gas? What if that gas is caustic and could be like acid on the moon-landing vehicle."

The man from Grumman was silent as he stood on the stage, and Darrell, without looking, could feel Les Myers shift in his chair. The blockhouse superintendent knew that Geisman was brilliant, but he wasn't sure if the big Texan was joking or voicing a valid concern.

"The dead moon theory is not something new," the Grumman man's voice sounded over theater. He was speaking into a microphone that made his words sound empty and metallic. "We do have the upcoming surveyor landing missions that will tell us about the environment our lunar landing party will have to face."

"So you think the moon is dead for a reason?" Les Geisman was like a bulldog with a bone. He grinned in the dark at the Grumman man's discomfort, but sat erect when Les Myers leaned forward with *the look*.

"Grumman doesn't know why the moon is dead," the engineer was shading his eyes from the projected moon image and peering toward the sound of the Texas accent, "but we're going to build a landing vehicle to find out."

"Now that's the attitude I like," Giesman said loud enough for everyone in the theater to hear, "even if it is from a New Yorker!"

Chapter 73

"God has no intention of setting a limit to the efforts of man to conquer space."

− *Pope XII*

December 4, 1965 - Cape Kennedy

"At least they'll be sleeping together," Les Geisman whispered from the corner of his mouth.

It was true Darrell thought as he smiled. On this Gemini flight, the two astronauts *were* scheduled for sleeping at the same time. On the earlier flights, real sleep was impossible as one pilot was awake with the other *trying* to sleep only inches away. Shoulder to shoulder, the Gemini spacecraft was cramped quarters. It was like a very advanced jet fighter cockpit for two, but this cockpit was flying at 18,000 miles per hour.

"Guido is Go Flight," Darrell said officially as he thought of the two men in the capsule and watched the projected flight path on his monitor.

"Fido is Go Flight," Geisman pitched his voice. The status notifications were rolling down the trench. It was the regular routine. There were vocal flight status reports every fifteen minutes.

When he was under the spotlight, Les Geisman was official, professional, and concise. He was as always, however, offering private comments and cracking jokes. Manned space flight, Darrell considered, was beginning to become routine.

The launch of Gemini-7 was a beautiful event. The rocket with the bright blue flame had gone up perfectly. The big Titan had once again performed flawlessly and placed Jim Lovell and Frank Borman into a low circular orbit. The two men in Gemini-7 were now circling above the earth and traveling alongside the second stage of the Titan booster that had propelled them into space.

As the control room settled down for a marathon flight and the Pan Am girls began moving down the rows of engineers at their monitors, the intercom crackled into life with the voice of an astronaut. The communication from high above the earth was relayed to Houston and the Cape from a series of tracking ships and remote ground stations across the globe. Sometimes the reception was good and sometimes it was not. The communication problem was obvious and everyone knew that with the upcoming Apollo moon missions, the current radio link was not going to work. The distance would be far too great.

"Gemini-7 here," Frank Borman was reporting through crackling static. "Houston, how do you read?"

After the launch, vocal communication was relayed to the Manned Space Flight center in Houston, Texas. Everyone at the Cape monitored their stations, but the voice of Capcom in Houston was the new vocal link to manned spacecraft orbiting the earth. The UHF reception from the Gemini missions was scratchy at best and everyone in Florida was listening carefully.

In contrast to the garbled transmission from the orbiting spacecraft, the voice from Houston was strong over the speakers. "Loud and clear — *Seven* — go ahead."

"We have a bogey at ten o'clock high," Borman's voice was determined and professional, but he also sounded slightly amazed. Through the electrical static, he was reporting an unidentified flying object in space.

"This is Houston . . . say again *Seven*?" Houston sounded skeptical.

"We have a bogey at ten o'clock high."

Heads began turning in the trench. Everyone was looking around and the control room was suddenly galvanized.

"Bogey" was a military term that usually meant unidentified and most often enemy. As Darrell and Les Geisman turned to look at each other, Kurt Debus and Wernher von Braun stopped pacing.

"Roger Gemini-7," the official tone from Texas continued to sound doubtful. "Is that the booster or is that an actual sighting?"

Borman's voice was quick to answer, "We have several," he reported. "Looks like debris up here. Actual sighting."

"You have any more information?" the voice from Houston was in a completely neutral tone. "Estimate distance and speed? Understand you also have booster in sight. Roger?"

"Yeah, have a very . . . very many," Borman hesitated, and it was easy to imagine him calculating. "Looks like hundreds of little particles banked on the left, out at about 3 to 4 miles."

The rock-steady voice from Houston continued. "Understand you have many small particles going by on the left."

Borman's voice over the intercom was suddenly garbled and static-filled. "Looks . . . like . . . a path to the vehicle at 90 degrees."

"Roger," the tone from Houston was clear and unruffled. "Understand that they are 3 to 4 miles away."

"They are passed now," Borman sounded relieved. "They were in polar orbit."

After a moment of silence with everyone in the control room very quiet, the voice from Houston sounded carefully curious, "Were these particles in addition to the booster and the bogey at ten o'clock?"

Borman's reply was a grave, "Roger."

Suddenly, Jim Lovell's voice overrode the intercom. He sounded excited. "I have the booster on *my* side," the new voice from space reported. "It's a brilliant body in the sun, against a black background, with trillions of particles on it."

"Roger, what direction is it from you?" Houston was now apparently only interested in what Lovell had to say. It was clear to everyone at the Cape that Texas did not want to talk about UFO's; they wanted to talk about the spent Titan booster that was traveling alongside in the same orbit.

Any ham radio operator could listen in on the conversation from space and report astronauts seeing flying saucers to the newspapers.

Lovell reported: "It's about at my two-o'clock position."

"Does that mean it's ahead of you?" Houston was referring to the spent Titan booster.

Lovell finalized: "It's ahead of us at two-o'clock . . . slowly tumbling."

"Roger Seven," Houston replied, and then the conversation went into the problems of station keeping with the still orbiting Titan second stage.

Les Geisman leaned forward, "Those two are going to be hounded by that later," he whispered. "That booster was venting fuel and oxidizer. That's why it was tumbling. It's Newton's third law: 'For every action there is a reaction.' Borman must have seen frozen particles of propellant in a big cloud—maybe the sun shining on it. I can't believe he reported a UFO."

Darrell shook his head and then leaned back in his chair, "Remember John Glenn's fireflies?" he said. "These guys are seeing new things with every trip. We're breaking new ground with every flight."

The big Texan eased his shoulders and looked down the trench. "I still say it was the sun on a cluster of frozen fuel particles, but with the sun pouring through a mass of propellant, it could have looked like a lit-up spaceship from the Saturday afternoon movies."

"That sounds about right," Darrell conceded. "But everybody here sure stopped what they were doing to listen. I think everyone wanted to believe it."

It was true; everyone in the control room was only now falling back into the standard routine. Even von Braun and Kurt Debus had stopped in their tracks to focus on the UFO report. They looked very anxious and they were still whispering in German.

After a few minutes, the Pan Am girls were smiling again and pouring coffee. The flight surgeon lit a cigar. It was back to business as usual. The UFO was gone but certainly not forgotten.

"Ready for the perigee burn?" Darrell asked.

"All set," Geisman replied. "Everything looks real good. I think we're going to make it—the whole two weeks in orbit. The Russians are going to hate this."

"For a minute," Darrell lowered his voice, "I thought that the bogey might be something the Communists sent up to stop us. I wouldn't put it past them. They know the world is watching and they sure do want to win."

Les grunted. "I thought about that too," he said, "and for a minute, before whatever it was disappeared into polar orbit, it sounded like Frank and Jim had it figured the same way." The big Texan shook his head and then turned back to his monitor. It was time to get busy.

There was an upcoming orbital maneuver burn that would propel Gemini-7 up and away from the low orbiting Titan booster. This would prepare the astronauts for the rendezvous with Gemini-6 that would be launched in eight days. Before the burn, however, the tumbling and venting booster with the huge cloud of sun-touched propellant crystals would be a ghost for Borman and Lovell and would haunt the two astronauts on several revolutions around the earth.

Whenever the two pilots expected to find the orbiting booster, it was not there. When they expected to be free and clear and well away from the second stage of the Titan rocket, the vast halo of sun-touched ice crystals would reappear as an unexpected apparition, and the spent fuselage would tumble through an orbital sunrise.

As the two men traveled through space, they were learning a new lesson. Maneuvering a spacecraft was like doing something complicated while looking in a mirror. Objects looked the same as in training, but they definitely were not. It was also very strange to realize that the only way to look fully at the earth passing below was when you were upside down.

Chapter 74

"Once you get to earth orbit . . . you're halfway to anywhere in the solar system."

— Robert Heinlein

Sunday Morning, December 12, 1965 - Cape Kennedy

"The timing should be perfect," Darrell explained to everyone in the trench. "When Gemini-7 crosses overhead, Jim and Frank should be able to watch the launch of Gemini-6—providing there's no cloud cover. The view should be incredible. "

The planning had been meticulous, and now, after 8 days in orbit, Frank Borman and Jim Lovell in Gemini-7 were speeding over California and headed for the east coast and Cape Kennedy at 18,000 miles an hour. The two pilots were 160 miles above the earth and waiting for the upcoming rendezvous in a near perfect circular orbit.

Meanwhile, the countdown clock on Gemini-6 was sweeping through the final seconds. Again the timing looked perfect. Les Geisman scratched his head and stared at his monitors. Once again everything on the Titan booster was showing nominal parameters, but as usual the operations crew in the blue Martin coveralls appeared anxious and apprehensive. There was nothing commonplace about lighting a rocket and launching men into space.

"We are . . . Five . . . Four . . . Three . . . we have engine start." The voice of the countdown was from Flight Director Chris Kraft.

High atop the latest Titan booster, Wally Shirra and Tom Stafford heard the voice of mission control and felt the twin engines of the big Titan start. At first, they thought they were on the way, but then only seconds after triggering the flight time clock, the Titan engines abruptly shut down.

The silence was deafening. No astronaut had ever been in a capsule atop an ignited rocket booster that had suddenly failed. When a rocket booster failed on the launching pad, a pressurized fuel explosion was usually immediate. It was the nightmare that every astronaut feared the most. *

With the flight time clock continuing to spin . . . in the ominous silence . . . Schirra and Stafford were completely focused on their previous training and on the abort –D handle lever that Schirra now held in his hand. The rigorous training was specific. Upon booster failure: eject — emergency egress.

For only a second Wally thought about the crushed heads of the crash test dummies that had not made it outside the compact Gemini cockpit. Then he thought of Ed White's space walk and how difficult it was for him to open and close the Gemini hatch. He then thought of Geunter Wendt and the big smile the larger-than-life German had given when he and his White Room crew had closed and sealed the two pilots into Gemini-6. Wally still held the lever for the ejection seats, but his intuition told him to hold fast. He felt no tremors or tale-tell signs of destruction or booster explosion — only the odd silence of engine shut down. The seconds dragged on. Tom Stafford was rock steady and listening for trouble as the two pilots sat shoulder to shoulder.

In the control room, Les Geisman said, "I am showing shut down but stable — all across the board. I believe they're all right. Tell them not to eject."

Over the intercom several voices were suddenly speaking at once. The voice that Darrell heard was that of Frank Borman crossing overhead in Gemini-7. All across the Florida peninsula it was a cloudless December sky, a peaceful Sunday morning, and the view from space was perfect.

"I saw ignition . . ." Frank reported from the ultimate vantage point, but his words were clearly disappointed. "And then shut down."

With a delicate silence enveloping the control room and visions of exploding boosters vivid in almost everyone's imagination, the calm, cool, and controlled voice of Wally Schirra came over the intercom.

"Fuel pressure is lowering," Wally reported carefully over the speakers. "We're below the redline values."

Les Geisman looked up. Everyone was watching. "The Martin crew are confirming," he nodded. "Fuel pressure is down. Wally's right."

Thirty minutes later the erector tower went up, and Guenter Wendt and his White Room crew in white coveralls were opening the hatches on Gemini-6. Guenter was very concerned when he personally opened the hatch, but he was beaming when his boys were safe and out of the capsule. Everyone knew that Wally Shirra and Tom Stafford were heroes for not ejecting. They had saved the mission.

The American astronaut program required that the pilots have a genius level IQ and be in perfect physical condition, but there was no way to test for nerves of steel and outstanding bravery. Tom Stafford and Wally Schirra had just passed the ultimate test.

In the control room von Braun and Kurt Debus were heads down and pacing. The engine shut down would take hours if not days to reconcile the problem. Meanwhile, Gemini-7 was on its eighth day of orbiting the earth and Gemini-6 was in trouble.

The Titan-2 Gemini booster had shut down because of the faulty release of an electrical umbilical plug. After only seconds of engine start, a clock in the Gemini capsule detected a malfunction and shut down the fuel flow from the big Titan's twin engines.

Chapter 75

*" . . . If we become extinct because we don't have a space program
. . . it will serve us right."*

Larry Niven, quoted by Arthur Clark

Wednesday Morning, December 15, 1965 - Cape Kennedy

"For the third time, go!" Wally Schirra's voice sounded over the control room intercom.

"Thank God for that," Darrell said over the rumbling thunder.

Gemini-6 was now on the way. It had taken three days and endless tests from the Martin crew in the blue coveralls, but the Titan booster was rising. It was almost clear of the gantry tower and the bright blue of the exhaust flame was already well above Launch Complex #19.

"Whoa!" Les Geisman said suddenly with alarm. "Did you see that shimmy?" Les was watching the flight director's screen. Without warning there seemed to be an abrupt and abnormal vibration as the Titan booster was rising.

Darrell did not have time to comment. His full attention was on the actual flight path as it began to materialize on his IBM monitor. The Cowboy had two flight paths to monitor: the actual and the theoretical. If the actual trajectory didn't line up beside the theoretical indicated on the screen, the pilots and the mission could be lost.

"Okay," Geisman continued carefully, "they seemed to have settled down."

After the brief vibration and blur on the television screen, the Martin-made Titan climbed effortlessly until fire-in-the-hole staging. With most rocket boosters, the second stage ignited only after the first stage shut down, but with Martin's Titan, the second stage ignited inside the first and produced a bath of orange flare light for the windows in the Gemini capsule. After the explosive burst of second-stage ignition, the astronauts experienced an incredible kick-in-the-butt acceleration when the second stage separated and powered away from the first.

When the Martin second stage shut down, Les Geisman leaned back in his chair. "This stress is killing me," he said. The big Texan then reached for his crumpled pack of Chesterfields and shook out a cigarette. He lit up and thoughtfully blew out a smoke ring

Darrell glanced at his monitor and completed a quick calculation. Gemini-6 at second-stage booster shut down was traveling at 17,600 miles per hour. Schirra and Stafford were well on their way. The unfolding flight plan was perfect.

"Flight path is right down the middle," Darrell offered officially. Von Braun was standing behind the Cowboy and looking carefully at the two flight paths on the flickering monitor. The German nodded, smiled, and after a quick clap on Darrell's shoulder, the Boss turned and continued down the trench.

"I know, I've always praised Martin and the Titans," Les lowered his voice, "but I'm glad this one is over. This booster was trouble from the start."

Darrell was still watching his monitor. Gemini-6 was now officially in orbit, but Wally and Tom were still trailing Borman and Lovell in Gemini-7 by 1300 nautical miles.

"Now why," Darrell asked as he worked the slide on the old K&E, "did you think there would be any more trouble?" Darrell was checking a new calculation but his mind was also on Geisman. Les never complained about stress and he did not offer negative comments—at least not seriously.

"Because of what they found after the loose electrical cable," Les leaned back and blew out another smoke ring.

He was looking around the control room. The Pan Am girls were on the move and the flight surgeon was lighting a fresh cigar.

Unofficially, after booster shutdown, it was time to catch your breath, thank God that the physics had actually worked, and take a few moments to reflect on the speed in which human beings had just been propelled into outer space. It was normal after booster shut down that everyone in the trench relaxed — at least for a moment.

After a Titan booster was launched, the pilots were finished riding 160 tons of volatile fuel and oxidizer, and as much as training and experience took charge, there was still the human factor and the possibility that one of thousands of components had been faulty . . . or tampered with. The Communist Russians hated the Gemini program and everyone at the Cape knew there had to be spies and espionage. There had also never been as many contractors or multinational technicians roaming the Cape. IBM alone had a survey crew of thirty on a daily basis.

"Okay," Darrell offered, "I'll bite. What did the Martin crew find besides the loose umbilical cable? By the way . . ." the Cowboy suddenly recalled, "that electrical cable . . ." He paused. "That was almost exactly what happened with Mercury Redstone-One."

"They'll start safety wiring those cables from now on," Geisman stretched in his chair. "Do you want to hear about the Titan trouble or not?"

"Sure Les," Darrell said soothingly. "Tell me what you know — especially now that it's over — tell me about the trouble."

Earlier when he arrived at the Cape, Darrell had seen the big Texan towering over two of the Martin technicians in the blue coveralls. The conversation had taken place under the spotlights by the gantry tower and just before sunrise. Everyone knew that Les always arrived early on launch days and always liked to walk up beside whatever mission was being launched. There was something special about the predawn bustle and the fueling of a rocket booster. Knowing that standing in the spotlights alongside the tower was the most complicated machine on the planet was always a moving experience and always inspiring.

"Tell me about the trouble," Darrell coaxed his friend. "Tell me about what you found out from the boys in blue."

"After the booster shutdown," Les began carefully, "when everyone was wondering and worrying about the ejection seats, there was a tale-tell warning sign that came up in the thrust-trace data. Even though the engine burn only lasted seconds, there was something on the graph paper that the Martin boys didn't like."

Geisman scowled and then paused to crack his knuckles. Again, this was unusual. Les was normally full of good humor and jokes, but today, Darrell could tell, the big Texan had really been under pressure. He had been worried like never before.

Over the intercom speakers, there were now four voices coming in from space.

"We copy Gemini-6 that you guys are finally up and on the way." Jim Lovell in Gemini-7 sounded excited. "We just saw your booster contrail as we were coming up and over Hawaii."

Frank Borman broke in jovially, "We are putting on our suits and waiting for company. It's really good to see you out there Gemini-6."

"Roger that Gemini-7 . . ." Tom Stafford acknowledged. "Third time is the charm," Wally Schirra added through the static.

Geisman had stopped to listen, but when the crackling transmission passed through a loss of signal zone, he went back to his worries over the troubled Titan.

"I can't help but wonder," he continued quietly, "what might have happened if that umbilical had not pulled out and shut down the booster when it did. The thrust-trace data showed a perfect engine start, but then there were some irregular lines appearing on the paper. The graph showed a definite drop in thrust — a dangerous drop."

"Are you talking about a booster failure that would have happened if the electrical cable had not pulled out?" Darrell was not focused on his side rule or monitor any longer. He was staring at Les Geisman with disbelief.

"There was just too much going on too fast—too much of a tight schedule." Les shook his head, "Apparently, someone left a dustcover on a gas generator. That's a critical engine part. It was hidden so deep that it took all night at pad #19 to find the problem. If someone hadn't noticed that thrust decay on the graph, and if that dustcover had not been found and removed . . ."

Once again, Geisman looked around the control room and shook his head. The Pan Am girls were pouring coffee. Kurt Debus was speaking German to a newcomer, and Les Meyers was standing with his hands on his hips and watching the flight director's television.

Walter Cronkite was on the screen. Behind the newscaster's desk was a futuristic backdrop featuring a sea of stars and a television version of outer space. On the front of the popular anchorman's console was a countdown clock showing the elapsed flight time for Gemini-6. The counting clock was scrolling through: thirteen minutes and fifteen seconds . . . and clicking away . . .

Darrell's thoughtful gaze was still a stare. "You're saying Les," he said very carefully, "that your predictable Titan could have fallen back and blown its self to pieces after that dustcover prevented enough thrust to get off the pad."

"That's about it," the Texas drawl agreed. "Of course, this is all need-to-know . . . but I'm worried that the dust cover might not have been an accident. I'm also worried that there might be more trouble."

Darrell had not moved since Geisman broached the subject. He suddenly thought of Mr. Smith and Mr. Jones from the Department of Defense. This was almost like spies against spies. After the premature release of the electric umbilical on Mercury Redstone-One this should have never happened again—unless it was no accident. NASA learned from every mistake and mistakes with rocket boosters were never repeated—ever.

On the television screen Walter Cronkite was holding a model of the black Gemini Capsule with the white service module collar attached. Even on the Gemini model the white letters on the black spacecraft were easily visible: "United States."

"*Good morning everyone,*" Cronkite's voice began over the television speakers. "*This ought to be . . . our most exciting day in space. Perhaps exceeded only by the very first of spaceflight with Alan Shepard's sub-orbital flight. For this is a day, when Wally Schirra and Tom Stafford . . . will meet and rendezvous with another spacecraft. It is a day when Frank Borman and Jim Lovell . . .*"

Chapter 76

"Demoralize the enemy from within by surprise, terror, sabotage, and assassination. This is the war of the future."

— Adolf Hitler

"Do you really think it was sabotage?" Darrell asked quietly. Les Geisman and the Cowboy were outside the blockhouse and standing under a cloudless December sky. In the distance, the Atlantic waves were breaking on the beach and the winter weather was as perfect as a Florida postcard.

Meanwhile, somewhere over the green jungles of Malaysia, and then abruptly over the deep blue Pacific, Wally Schirra and Tom Stafford in Gemini-6 were far above the atmosphere and streaking toward the planned rendezvous with the bearded astronauts Borman and Lovell in Gemini-7. Gemini-7 had been orbiting the globe for 11 days and was rapidly headed for a new record in miles traveled in outer space.

"Imagine this as a set-up." Les looked thoughtful as he turned from the ocean to examine the orange erector tower of Gemini's pad #19. "What if our spies—that are the good guys—found out there was something wrong with the booster. What if our guys learned that something had been tampered with but they didn't know what it was? What if our guys fixed the umbilical to come loose and stop the launch before whatever that had been tampered with could go wrong?"

Darrell considered the scenario. A cable had indeed parted prematurely on the first unmanned Mercury and

289

that misfired missile had been painstakingly analyzed. If the umbilical cable had been rigged to cause a premature but safe engine shutdown, whoever rigged the cable must have known that an immediate inspection of all the Titan systems would follow. If Geisman's scenario played out as outlined, it would reveal any real and dangerous sabotage that was already in place on the booster.

Calling Washington, the people at Martin, or even the brass at NASA might not have been enough to warrant a flight cancelation of the most important mission to date. It was a lot to think about — spies verses spies — but with Les's theory the equation *could* add up to espionage and counter-espionage and perhaps even sabotage.

If the mysterious dustcover had blocked the gas generator and the Titan fell back and exploded, there would be no rendezvous in space and Kennedy's plan for a moon landing could certainly have been delayed. Wally and Tom might have been killed and there would have been a dangerous drop in American morale and confidence throughout the free world.

No one would have known there was a dustcover problem without the failed umbilical that had already happened before on the first Mercury Redstone. NASA did not repeat mistakes and the simple fact that another umbilical had pulled out was alarming. It was also disturbing that the Martin technicians had missed the dustcover. Geisman's scenario was certainly much more complicated than the formula for any spy movie, but this was a formula that might very well be real.

For the first time since Darrell had known the big Texan, Les Geisman was showing real concern. He was not in his normal lighthearted mood, and he was definitely not joking. It was a wild idea — right out of an Alfred Hitchcock movie — but as the Cowboy looked out over the whitecaps on the ocean and the orange gantry tower on pad #19, he could not discount the possibilities. In the distance, there were figures moving at the base of the orange gantry wearing the blue Martin coveralls.

"I'm glad no one is listening to us talk about this," Darrell offered quietly. "It just sounds crazy." He then cracked a smile as he thought of Will Hammond's wild tales of the desert out at White Sands. His concern darkened however when he considered that one of the technicians in the blue Martin gear might be much more than he appeared.

Les didn't see the smile, but he did see the frown and the expression of concern. "There's a lot to think about and more than one thing for certain," Geisman paused to blow out a smoke ring and then held up finger number one.

"If that dustcover had remained in place, Gemini-6 would have burned bright for a few seconds and then fallen back and exploded on almost every television screen around the world."

Finger number two came up: "If that cable had not parted, the dustcover on the gas generator would have never been found and we go back to the explosion on Pad #19."

When finger number three came up, Giesman shook his head and then his forehead wrinkled with concern. "Whenever I come across a problem with a lot of possibilities, I always remember something I learned down at the University of Texas: *If there are a number of explanations for observed phenomena, the simplest explanation is preferred.*"

Before Darrell could comment, Les dropped all the fingers but inclined his head toward the erector gantry tower and the concrete structure that was pad #19. Even with two Gemini missions up at the same time and the Gemini-8 launch date weeks away, the launching pad beside the waves breaking on the beach was a bustle of activity — the McDonnell-Gemini technicians in white coveralls and the Martin-Titan crew in blue.

"Suddenly and without any previous trouble," Geisman offered darkly, "we have two major problems on a single rocket booster that, before all this mess, was the most reliable booster and ballistic missile ever made. Now I ask you, what is the simplest explanation for this very complicated scenario?"

When Darrell and Les entered the blockhouse and took their places in the trench, everything and everyone looked a little different, but the excited voices from space still sounded the same. The buoyant voices were on a grand adventure, streaking above the earth, and it was hard to imagine a dark force on the other side of the globe that didn't want that adventure to happen.

Chapter 77

"When a man looks across a street, sees a pretty girl, and waves at her, that's not a rendezvous, that's an acquaintance. When he walks across the street and nibbles on her ear, that's a rendez-vous!"

— *Wally Schirra*

After 86 minutes in orbit, Gemini-6 was riding high and coming up over New Orleans. With a quick look out his viewport window, Wally Schirra gripped the flight control stick and ignited the spacecraft thrusters. When the boosting adjustment burn was complete, the two men in the McDonnell spacecraft were traveling four meters per second faster. Now, as they shot around the globe, their low trajectory altitude would remain the same, but the height of their new orbit would be over 160 miles high. Because Gemini-6 was still closer to the earth, they were going faster and gaining on Borman and Lovell in Gemini-7.

When the two travelers in Gemini-6 were crossing over Diego Garcia, and headed for Australia, Wally took the stick again and performed another precise maneuver. It was exactly 52 minutes after the speed adjustment burn was made over New Orleans. This time the goal was to increase the perigee speed or point of low orbit progress another 19 meters per second faster. This was a precise maneuver selected by the onboard computer and would place Gemini-6 on the same orbital plane as Borman and Lovell in Gemini-7.

"Wally, it looks like we're right down the middle," Tom Stafford reported after looking at the flight plan and the elapsed flight time clock on the central control panel. The flight plan was thick as a telephone book, but after the clock was turning through three hours since launch, Gemini-6 was right where they were supposed to be—almost to the second.

"Yeah, it's looking really good Tom," Wally said, "coming up on AOS."

AOS was acquisition of signal. The Carnarvon tracking station had replaced the original Muchea-Mercury site and was now the Gemini program's voice down under. Gemini-6 had already passed down range from one of the South Seas tracking ships, and the next upcoming antenna array was the new broadcasting station in Australia. AOS was the reassuring routine when the orbiting astronauts came in receiving range of a ground-based radio transmitter. LOS was loss of signal and happened when a spacecraft was out of range of a ground station's receiving ability. When the pilots were traveling overhead and through a corridor where ground support communication was unavailable, they were in LOS. Loss of signal was a lonely place in orbit until the streaking capsule came in contact with a new and friendly transmitter. As everyone knew, there were many ground- and sea-based radio stations that were not friendly to any spacecraft labeled with the letters "United States."

"M equals four," Les Geisman offered when the vocal communication dropped off after the South Seas LOS.

"I heard about that one," Darrell's response was preoccupied. He was keeping track of two Gemini spacecraft at once. "You know that 'M' doesn't mean anything in that equation and four only means that the rendezvous is scheduled for the fourth orbit."

"I know," Geisman shook out another Chesterfield. "It looks like the Martin people are out of the woods but not the folks from McDonnell. The booster made it but what if there's something wrong with the spacecraft itself?"

"Do Wally and Tom know about the dustcover on the Titan and your theory on the electrical umbilical?"

"You bet." Les paused to smile at one of the girls from Pan Am as she arrived to pour coffee. No one in the trench would ever voice any worries or concerns within earshot of the girls from Pan Am. When the flight attendants were nearby, everything was perfect.

"We have AOS," the voice of Elliot See the Cap-com link broke over the control room speakers.

With the sandy Australian Outback passing 160 miles below, Wally gripped the control stick again to start an orbital phase adjustment burn. He fired the aft thrusters, and he and Tom Stafford felt Gemini-6 surge forward. Outside the spacecraft windows, the full moon was coming up over the earth's horizon and had never looked closer. From space and beyond the atmosphere, the details on the lunar face were beyond inspiring. The ancient and crumbling craters were everywhere, and the jagged mountains between the gray oceans of dust very distinct. Below the spacecraft and under the earthbound clouds, the vastness of the Pacific Ocean was reaching out toward South America.

"All the thrusters seem to be working fine and guidance is right down the middle" Darrell closed his K&E. "They just picked up another 62 feet per second."

"Yes sir," Les sipped his coffee, "but the tricky part is still on the way. Remember the Russians tried this and the closest they could get was a few miles apart. They saw each other go by — like a couple of bullets shot out of different guns — but that was about it. We really can't imagine what it's like trying to chase and close in on another object that's orbiting and traveling at 18,000 miles an hour. This rendezvous is ground breaking stuff — but if it doesn't work — we're not going to the moon."

Geisman then sat back in his chair in simulation of a pilot in space. He looked forward and then down. "Try to picture a tiny little target flying through orbital space," he said, "with the oceans, the clouds, and the continents turning below . . . Just thinking about what it must look like makes me seasick."

Les shook his head and refocused on his monitor. It was time to go back to work. He was concerned about a misbehaving fuel cell in Gemini-7. The big Texan then stretched, resettled in his chair, and began rechecking a calculation on his slide rule. The chores in the trench never stopped.

Despite Giesman's uncharacteristic worries and the troubles over the Titan booster, Darrell was impressed. Wally and Tom were flying Gemini-6 and following the theoretical fight path almost to the letter. Wally, of course, was following the directions of the IBM computer, and Tom Stafford was clearing the board and constantly punching in new numbers, but the results were right down the middle. Everything was going perfectly except for the nagging thoughts of sabotage and what could have happened to the Gemini-6 Titan booster if the mysterious dustcover had not been found.

Wally's voice sounded distant as it crackled over the speakers, "Plane adjustment burn complete."

"He's turned south," Darrell was watching his monitor. He then rattled off a quick list of figures and Les began a new calculation on his K&E.

The Texan was concerned but he was also excited. Darrell needed help and the fuel cell worries could wait.

Elliott See was speaking from Houston and his voice was perfectly clear over the blockhouse intercom. "We show slant range distance at 198 nautical miles. You guys are looking really good Gemini-6."

"Roger Houston," the speakers crackled, "we copy slant range distance is right down the middle."

"That last adjustment burn kicked it in the butt." Geisman had finished his calculation and for the first time since the Gemini-6 liftoff, he sounded like his old self. "They're catching up quick," he said. "The slant range is really closer to 150 miles."

"Gemini-6," Houston coaxed. "You should start to have *Seven* as a radar image."

Wally responded through the static. "Roger that Houston. Tom saw a radar flicker and now we have a good solid lock. There's no doubt about it. We have a hard radar lock on something. It *must* be *Seven*."

"They're almost at a perfect circular orbit and they're gaining fast." Darrell was comparing the actual flight path of the fast approaching "6" to the stationary orbit of Gemini-7. The Cowboy had been concentrating on his monitors, and he knew Giesman was helping and close on his right, but when he looked up, he realized von Braun, Kurt Debus, Les Myers, and two others from behind the trench were now crowding in and looking over his shoulder. "Wally and Tom are coming up from underneath," he said, "and they're really moving."

"Oh my God," Wally Schirra's voice broke over the speakers with a fresh crackle of static. "There is a real bright star out there. It must be Sirius. I'm diming the navigation lights on my side to see it better."

"Bring us a star chart," Les Myers sent the order across the control room and a youthful technician fled with the command.

Les Giesman grunted. "That's not Sirius they're seeing," he offered. "That's Gemini-7 in the stationary orbit. I'll bet a paycheck on it."

Without comment, Les Myers picked up the phone and began speaking quietly.

"Of course," von Braun acknowledged, "it's the sun shining on Gemini-7. There can be no mistake. Gemini-6 is traveling through nightfall and they are already below the sunset. Wally is just excited. He is seeing an artificial star."

Kurt Debus voiced his agreement, "Ja, Ja, it can only be the other spacecraft . . . another spacecraft in the sunlight and above the darkness and the horizon." Debus' English was heavily accented but it was getting better every day.

"That sounds about right," Darrell confirmed. "*Six* is now only 62 miles below *Seven* and coming up very quick. With the sunshine on the metal that must be what they're seeing."

Elliot See's voice broke over the speakers just after Les Myers hung up the phone to Texas. "*Six*," the Cap-com voice from Houston began, "we believe that you are actually seeing *Seven* and not a star."

Les Myers nodded with satisfaction. President Lyndon Johnson was all about fortifying the space program in his native Texas, but Myers had just explained long distance to Houston what the rocket boys at the Florida Cape had already figured out.

"Roger that Houston," Wally sounded fuzzy from beyond the atmosphere. "We copy that our Sirius is actually *Seven* and radar is showing they are now passing into nightfall."

As Wally Schirra scanned the edge of darkness that was a moving sunset racing across the earth, Tom Stafford suddenly picked out the flashing navigation lights from Gemini-7. When the glare of the sun was no longer searing above the earth's horizon, Sirius was abruptly replaced, and the McDonnell spacecraft in the higher orbit was easily visible. Without hesitation Wally tilted the control stick toward the pulsing beacons and pushed the thruster button. Gemini-6 responded with a delicate rush of power and the gap between the two American spacecraft closed tighter.

After another twelve minutes, with the dark Mediterranean Sea and the tiny lights of the Greek islands passing below, Wally coaxed the stick and the thrusters again, and the distance between the two spacecraft narrowed dramatically. Gemini-6 was now half a mile away and still closing by over a mile a minute. Ironically, the two spacecraft with the block letters "United States" were just passing over Turkey, but they could see into southern Russia and Kazakhstan. They were almost over the Soviet rocket launching facility with their navigation lights probing into the Communist midnight.

The two spacecraft were now only a thousand feet apart with the moon-lit clouds, the continents, and the mountains of the earth passing below. Behind and below was Mount Ararat and rumored resting place of Noah's ark.

From the edge of the trench the technician that had gone to fetch the star chart was back. He was a very thin and gangly young man with a white shirt, a black tie, and black horn-rimmed eyeglasses. The astronomy chart in his hands was big, unrolled, and cumbersome. "Sirius is not out there," he said as the paper rattled. "Not where they're looking, but they should be seeing the Gemini stars any minute: Castor and Pollux."

The young astronomer was crestfallen when no one acknowledged his finding, but he beamed like a peacock when the intercom crackled back into life with a fresh burst of static.

"Our view looks quite fitting, Gemini-7," Tom Stafford reported from his side of Gemini-6. "We can see the Gemini constellation just above your position. That's the Gemini stars Castor and Pollux. No doubt about it. The view is inspiring."

As Tom Stafford transmitted, Wally pulled the flight control stick back and hit the reverse thrusters. It was time to slow down, and the two men instantly felt the brakes.

"Wow! Gemini-6, that looked incredible," Frank Borman reported excitedly as he watched the sister spacecraft approach. "That was some fireworks. Wally, your retro burn was a tongue of flame that must have shot out for over forty feet. That was impressive. You don't see that in a simulator."

"They're only about 130 feet apart," Darrell looked up from the actual flight path on the monitor. "They're coasting at exactly the same speed. They've done it. They're traveling at 18,000 miles an hour and they're getting closer. They've figured it out. They really are space pilots now."

"Ja, Ja, very good," von Braun clapped Darrell on the shoulder, "but it is all of us together — all of us working together that make this miracle happen."

"We show rendezvous, Gemini-6 and 7," Elliott See from Houston reported. "We're celebrating here and lighting cigars. We have celebrations starting here and at the Cape. We show that on December 15, 1965, at 2:33 p.m., that Gemini-6 has officially rendezvoused with Gemini-7. That's history — American history. "

Before anyone could do anything but click open Zippos and start lighting cigarettes, the Pan Am girls were on the move and placing American flags with little stands on top of all the monitors. The girls were moving in front of the rows alongside Les Myers. He was beaming as he helped with the American flags. The flight surgeon lit a cigar. Cheers erupted all across the control room, but in the front row and in the trench the tension was still intense.

Les Geisman shook out Chesterfields, and the Cowboy from Iowa, the big bear from West Texas, and the German rocket scientists that had changed the world lit up. The mission was far from over but it was another step toward the moon and the not-so-distant lunar surface.

Chapter 78

"Gemini was a tough go. It was smaller than the front seat of a Volkswagen bug. It made Apollo seem like a super-duper, plush touring bus."

— *Frank Borman*

160 Miles Above Iceland

"Roger Houston. We copy that you copy, and the rendezvous is history," Borman responded, but he sounded troubled and preoccupied. His transmission then shifted back to Stafford and Schirra in Gemini-6.

Despite the scratchy radio reception, it was suddenly very clear to anyone listening that the Space Flight Center in Houston and everyone at the Cape were now on the sidelines. There was concern mixed with the static in the voices that were coming from high above the earth because there was something wrong with Gemini-6. Something was broken off at the rear of the spacecraft and trailing through space.

"You've got a lot of stuff all around the back end of you," Borman continued, and there was clearly tension in his voice. He sounded worried. "You're trailing what looks like cords or broken-off wires at about 15 to 20 feet. There is quite a lot of debris coming out from behind your service module."

Before any comment was possible, Darrell chanced a glance at Les Geisman. The big Texan was clearly concerned about the report of debris and broken-off wires. His forehead became a furrow of wrinkles before he bit his lip and shook his head. Broken parts streaming out from behind the tainted Gemini-6 was anything but good news.

Just as the two Gemini spacecraft were passing over Halifax and the unseen fishing fleets of the Canadian maritime, Wally Schirra saw that something was also wrong with Gemini-7. He then tipped the control lever gently, tapped the trust button, and rolled his spacecraft closer to his sister ship. Just before the roll was complete, he compensated with the yaw thrusters. Before the yaw maneuver was finished he fired another burst from his braking thruster and when the bright orange flame and the exhaust flare was finished, he and Tom Stafford were close enough to see that the cabin lights were on in Gemini-7. He could also easily distinguish the beard-stubbled faces of Frank Borman and Jim Lovell looking out of their perspective windows. One of the astronauts was cleaning the inside of his viewport with a white cloth. The two spacecraft were about twenty feet apart and completely controlled and stabilized. They were also racing above the earth with a groundspeed of 18,000 miles an hour.

Wally transmitted, and his voice at the Cape was reassuring even if it was filled with the crackling static. "So do you," he said as he spoke to the bearded Borman and Lovell, "you guys have what must be the same thing—lots of stringers trailing behind. The streamers look like some kind of plastic. It must be something left over from the Titan separation. I'll bet every mission has had the same leftover booster streamers but no one else has been up here to notice. We can't see what's trailing out behind us when we're looking forward."

When the voice of Elliott See broke in over the speakers, everyone at the Cape knew that Flight Director Chris Kraft was motivating the Gemini astronaut on the ground. Elliott See might be a fellow Gemini astronaut, the Cap-com link over the radio, but Chris Kraft had a way of broadcasting his command and influence even if his subordinates *were* in outer space. For the first time ever, the Cap-com link sounded uncertain over the intercom.

"Ah," Elliott See began, "how much maneuvering fuel do you have remaining Gemini-6?"

Darrell looked around the control room. Everyone had looked up. This was not a real question. Everyone knew that the constantly broadcast telemetry would easily reveal the maneuvering fuel levels to everyone at the Cape *and* to everyone at Houston. This was a transmission that really said: Flight Director Kraft is concerned that Wally and Tom might be burning up too much fuel as they fly around like a couple of barnstorming pilots showing off at a county fair.

Suddenly Wally Schirra was all business. He understood very well the meaning of the message. "We have over 62 percent remaining," he transmitted confidently. "As of five hours and fifty nine minutes into the mission we have used 113 pounds of fuel. I say again. I have over 62 percent remaining."

"Roger Gemini-6," Houston's reply once again sounded routine over the speakers. "We copy 113 pounds. You are 'Go' for station keeping."

"Roger Houston," Wally grinned as he finished the transmission. He then looked over to Tom Stafford and winked. Before Stafford could react, Schirra tilted the control lever forward and tapped the thruster button, and the stationary spacecraft that was Gemini-7 loomed closer.

"That is real impressive station keeping *Six*." Jim Lovell's voice sounded impressed over the speakers. "We estimate that you and Tom are about 9 feet away."

"Gemini-6," Elliot See was once again calling from Houston. "You are 'Go' for station keeping and formation flying. Gemini-7 you are to stop all rendezvous maneuvers when you reach 11 percent of maneuvering fuel remaining."

"Roger that Houston," Jim Lovell responded through the static.

Before there could be any more radio traffic, Wally tipped the control stick again and did a complete fly-around of Gemini-7. As the white block letters "United States" rotated in view on the black capsule, Schirra and Stafford could easily see Jim Lovell once again wiping the inside of his viewport window in Gemini-7.

"Congratulations Wally. That sure is some great flying," Frank Borman voiced the compliment.

Gemini-6 was now pointed at *Seven* and the snout of *Six* was stationary and pointed at the two bearded astronauts that were looking out of their windows. The nose of Gemini-6 was less than two feet from Gemini-7 and both spacecraft were completely stationary and stable. They were also crossing over the Atlantic Ocean and headed toward a new sunrise and a new day that, at orbital speed, would only last forty-five minutes. Watching a sunrise from space, however, was inspiration enough to last a lifetime.

When the dawn and the flash of rising sunlight suddenly touched the earthbound atmosphere, the edge of darkness vanished and was instantly aglow with every color in the aquamarine. Beneath the new day and the fresh breath of life were the blue marbled oceans and the vast and ranging clouds streaming between the continents.

When Schirra titled the lever and touched the yaw thruster, he and Stafford turned away. After the sunrise crested, the unbridled star of the solar system poured above the edge of the atmosphere and sent a blinding and unfiltered light into Wally's left viewport and then into Tom Stafford's. Just after the sunrise raced over the ocean below, Tom took the stick and easily steered "*Six*" over to another station-keeping position at about 30 feet away from where Gemini-7 rode as steady as a rock.

"This is going to be easy," Wally announced to everyone over the intercom. "This really proves it. If it's this easy to steer within one foot of another orbiting ship then docking should be no problem. No problem at all."

During the next forty-five minutes of sunshine the navigation lights on both spacecraft blinked easily even in the pouring light. When Borman and Lovell used a parcel of their allotted maneuvering fuel for a fly-around, even the smallest details of both spacecraft remained easily visible to each other. When it was time for "*Seven*" to look closely at "*Six*," Wally held up a sign that read: "Beat Army."

Wally was a Navy flier and both Lovell and Borman were Army pilots. Schirra was flashing a penlight through his viewport window to attract attention, and when he was certain the pilots in *Seven* saw the signal, he set the little flashlight to illuminate the lettering on his makeshift message. The highly competitive Army versus Navy football game was only days away.

When Houston called for an end to the rendezvous maneuvers, the two sister ships had made three complete obits together and were never separated by more than 300 feet. After the rendezvous excitement all the pilots were hungry. With Wally keeping an eye on the distance between Gemini-7 and his own ship, Tom Stafford prepared dinner.

The first course was rehydrated shrimp cocktail. The main course was chicken stew and vegetables. Apple juice was the beverage and chocolate pudding was desert. All the foods and drinks were pulled through a plastic tube — straw-style — to keep floating food particles away from vital tools and moisture sensitive instruments. When the meal was finished and the empty food packages tucked away, another orbital day was complete and once again the sun dipped below the earth horizon.

As the constantly moving veil of nightfall began to cover and cross into darkness, electrical lights from the nestled cities began to wink and glisten. The big cities were easy to identify but the little towns and villages were as mysterious as John Glenn's fireflies. Even with the most advanced education possible the travelers in space could only look down and guess at the names of the hamlets that were the tiny lights passing below.

That was the thing about space, Wally Schirra considered just before he drifted off to sleep: You never really knew what you were going to see but it was always different from what you expected.

Chapter 79

"In Mercury you couldn't translate. You could just change attitude. But you were actually flying it like a flying machine in Gemini. Gemini was just about the right size."

— *Wally Schirra*

After the tedious pre-launch preparations, the launch, six hours to maneuver and rendezvous, and five hours of fly-around and station keeping, everyone in the trench was as exhausted as the pilots that were circling the earth. Everyone in the control room had been at their stations for over 14 hours.

Darrell leaned back in his chair and lit a Chesterfield. The cigarette was from Les Giesman's pack. Darrell was out. As he smoked, he said to Geisman, "After Wally did the flip around they're riding at about 9 miles apart."

"Yeah, that should be good," Geisman nodded his agreement. "At least for watching the sunset. It really is hard to imagine that there are four men asleep and riding around the earth every ninety minutes."

"They're asleep alright," Dr. Charlie Berry offered. He was the flight surgeon and also stationed in the trench. "All their heartbeats and breathing show that everybody is asleep. And when my relief gets here I'm headed for bed myself."

"Did you guys hear what Borman said to the tracking ship just after they all finished eating and Houston ordered them to settle down for the night?"

When Les and Darrell both wearily shook their heads, Berry offered, "Borman said: 'We have company tonight.'"

"I don't know guys," the surgeon continued, "I'm worried about *Seven*. Two weeks up in that thing could be dangerous. Gordo Cooper could hardly stand up after only 33 orbits and all the medical people have been concerned ever since. I don't know . . . maybe I'm just tired. Maybe Kennedy's goal is too much. He should have never been riding in that open car. He was too good of a president to be killed like that — too much of a visionary."

Before anyone could comment, Berry's replacement arrived and the flight surgeon was away and into the Florida night.

"What do you think Darrell?" Geisman said when the surgeon's replacement had taken his seat and was out of earshot. "Do you think it's too much? Kennedy's goal and a man on the moon?"

"I think it's the greatest thing that's ever happened," the Cowboy responded softly. "But don't worry too much Les." Darrell used the classic phrase, "I think spies or not, we just might have made it through this one."

"Brother," Les Geisman said, "I hope you're right."

Chapter 80

"Be thankful for problems. If they were less difficult . . . someone with less ability might have your job."

— Jim Lovell

December 16, 1965

When reveille played, Les, Darrell, and all the others were back in the trench. It was now standard procedure to play the classic trumpeted call to wake the troops whenever there were military men that were asleep in space.

"Gemini-6 do you copy?" once again the voice from Houston was Elliott See. He sounded refreshed.

The speakers crackled. "We copy Houston." The voice just crossing over the Florida peninsula was Tom Stafford. He also sounded refreshed. For the men in space and everyone at the Cape it was still nightfall, but sunrise was only moments away. Both sets of the astronauts could see the predawn light racing across the Atlantic Ocean. The sunrise view from the window ports was like watching creation.

"Gemini-7 . . . this is Gemini 6 . . . do you copy?" Stafford suddenly sounded very excited over the intercom. Racing into the sunrise, the two spacecraft were easily visible to one another and traveling at about seven miles apart.

"We have sighted an object going from north to south," Stafford reported, "probably in polar orbit; it looks like he might be planning to reenter soon. I see a command module with eight smaller modules out in front. The pilot of the command module appears to be wearing a red exposure suit.

Standby one . . ." Stafford's voice crackled. "You just might be able to pick him up. I think we can hear something."

With everyone in the trench at first spellbound with the report of another UFO, but also just beginning to smile, the sounds of a distant harmonica broke in over the speakers along with the accompaniment of tiny ringing bells. At first, the sound was unclear, but soon the notes and the familiar melody of "Jingle Bells" began sounding over the speakers. It was only December 16, but for the moment, and for everyone listening, Christmas had arrived early. Christmas was coming in from outer space. All the girls from Pan Am were beaming. Les Myers grinned and then shook his head. Kurt Debus looked confused until von Braun began speaking rapidly in German.

With Houston still quiet and Cap-com offering no official response and everyone at the Cape wondering what the strict and by-the-book Flight Director Chris Kraft was going to do about harmonicas, jingle bells, and another prankish break in NASA protocol, Wally Schirra who had been playing the harmonica, spoke over the sporadic static. "Really good job Frank and Jim." Wally was now all business as he spoke to Lovell and Borman in Gemini-7. "We'll see you on the beach."

After Tom Stafford stowed away the jingle bells and Schirra tucked away the harmonica, Stafford jettisoned the white service module collar and Wally took the control stick and flipped Gemini-6 around. He then pitched the black capsule to fly "snout down" so the two pilots could watch the reentry zone as the sun-touched Atlantic raced into view.

Darrell shook his head, and then motioned for Les Myers. After the control room superintendent was summoned, Kurt Debus and von Braun immediately began heading for the trench. "Jingle Bells" was forgotten. When everyone was collected together and looking at the Cowboy's monitor, Darrell offered simply, "They've overshot the landing zone."

"How much are they off?" Les Geisman stubbed out his cigarette and leaned toward the guidance monitor, "How much are they over?"

"Too much," Darrell said. "They might have to make another orbit." The Cowboy tapped his slide rule. "If they don't do something quick, they'll miss the recovery ship."

Without comment, Les Myers was on the phone to Houston. The line to Texas was always open.

The Cowboy did not voice unwarranted objections. He did not raise false alarms. Before the conversation to Texas was complete, Schirra's voice sounded over the speakers. "We're a little over on the LZ," he reported through building static. "I'm going to bank her a little — back and forth. A little yaw will slow her down. We're just going to tap the brakes . . . just a little."

"Ach Zo," Kurt Debus exclaimed and sucked in his breath. With everyone watching the guidance monitor, Wally Schirra began doing the undoable. On the Cowboy's monitor, the Gemini-6 actual flight path was well ahead of the theoretical, but as the new telemetry began streaming in, it was obvious that Wally was indeed tapping the brakes. He was titling the capsule back and forth in the outer fringes of the atmosphere to slow down. He was air braking and Tom Stafford was doing the math. They were a team and they were flying a spacecraft like no one had ever done before.

As the huddle of engineers watched the actual flight path merge with the planned and perfect-case scenario, it was suddenly obvious that there was an amazing talent at work and a very special set of newly learned skills. Gemini-6 was back on track and right down the middle.

Wernher von Braun nodded, then he smiled, "I believe we are watching the results of a new kind of evolution." The German looked around as if his convictions might somehow be denied. "In just a few hours we have two sets of pilots that have truly learned how to fly a spacecraft. We have just seen the future, and the future is now."

When Schirra's and Stafford's braking maneuvers were complete, they dropped Gemini-6 out of the sky within view of the television cameras on the USS Wasp and television coverage was broadcast around the world. When the two astronauts in Gemini-6 splashed into the Atlantic waves, their spacecraft was only eight miles from the recovery ship. The McDonnell-Gemini spacecraft was an amazing machine and all the astronauts loved to fly her.

Gemini-7 with Frank Borman and Jim Lovell remained in orbit for their entire 14-day mission. They returned to earth two days after Tom Stafford and Wally Schirra on December 18, 1965.

Chapter 81

"Fly me to the moon . . . and let me play among the stars . . . let me see what spring is like on Jupiter and Mars . . ."

— Lyrics by Bart Howard and recorded and released by Dean Martin in 1966

When Bill Hinkle arrived, Darrell was surprised. He didn't know the Chrysler man was coming and he didn't know that Will Hammond from Huntsville would be traveling with him. Everyone did know that Chrysler, Boeing, and Douglas had the contracts for building the moon-ship boosters and NASA was the overseer of Wernher von Braun's brand new baby.

The two men had just arrived from New Orleans and were at the Cape to observe firsthand how the new vertical assembly building was taking shape—Bill Hinkle officially from Chrysler and Will Hammond from the new Saturn booster plant just outside New Orleans. For the latest age of rocket-patch soldiers, New Orleans was the new Huntsville. The Chrysler plant in Detroit and the Redstone facilities in Huntsville were not nearly big enough to build the massive first stage of the Saturn-5. The Saturn-5, when complete and stacked together, would be the largest and most complex rocket booster ever built.

"The new construction looks good Darrell—" Hinkle exclaimed, "real good—and the new blockhouse with the big windows, that's great."

"It looks like you boys have built quite the barn," Hammond agreed, "and launching pad and control rooms like nothing before. I guess we've come a long way from piling up sand with a bulldozer to make a blast-proof wall."

Darrell couldn't stop the smile and Les Geisman look puzzled. The Cowboy had never told the big Texan about the Redstone's only failure and the explosion at White Sands that the technicians were still talking about. In a flash, Darrell imagined Will Hammond's enhanced version of the Redstone guidance failure and his embellished story of the booster explosion at the Proving Grounds. Given Hammond's colorful character, he was probably telling the Redstone misfire story to all his new friends in Louisiana.

After Hinkle and Hammond's arrival, Darrell and Geisman were walking with the newcomers along the crushed gravel road toward the new launching pad that was under construction. At the opposite end of the three-mile road and the new launch complex was the recently completed vertical assembly building and the assemblage site of the new Saturn booster and the Apollo spacecraft.

The mammoth 52-story building was finished and the hurricane-proof corrugated siding was bright with new paint. The framework of the new assembly building was over five hundred feet tall, and the new structure was the largest building in the world by volume. It was so huge that on hot and humid days, condensation would form on the metal cross members and it would drip dew inside. When not climate controlled with 10,000 tons of air-conditioning, the largest building on the planet had its own interior weather. Painted on the side of the colossal construction was the circular NASA insignia with the blue sphere, the white stars, and the red vector piercing a circular orbit. It was the same emblem patch that was stitched over the breast pocket of Will Hammond's white, starched, and perfectly ironed uniform shirt.

"The first of the Saturn-5s are already well under construction," Hinkle offered when the four men had found a stopping point midway on the road. It was late afternoon, summer, and a thunderstorm was building in the distance. Lightning spiked out over the ocean, and thunder rumbled and echoed off the mammoth rocket hanger.

When the thunder fell silent, Will Hammond and Bill Hinkle were looking around, first to the new and massive assembly building and then over to the newest gantry tower and elevated concrete launching pad. The new vertical assembly building was gargantuan and monolithic and in sharp contrast to the flat Florida landscape. The enormous new launch complex and the new blockhouse with the big slanting windows was in every way futuristic. Even the crushed-rock roadbed between the assembly building and the Saturn launch site was impressive, as was the largest treaded crawler ever built. The enormous flat-deck transport crawler would creep forward at only a snail's pace, but it would carry the weight of a moon ship that would be 40 stories tall.

Les Geisman was looking across the distance to pad #19 and the orange Gemini erector tower. At pad #19, there was no current activity, no Martin technicians in blue coveralls, and no Titan booster waiting to be launched. Project Gemini was over. The Titan two-stage would always be a great workhorse for placing satellites in orbit, but now it was time for the biggest rocket ever to lift three astronauts and fifty tons of cargo to the moon.

At the new Apollo pad labeled #39-A, there was a string of concrete trucks waiting in line with their holding vessels turning. Pad #39-A was still under construction and the concrete trucks seemed endless. Beyond the construction and the beach, the Atlantic Ocean was rough. The wind from the thunderstorm was gaining strength and the whitecaps on the water were fierce. The approaching storm was rapidly cooling the heat of the late afternoon.

"Tell us about the new Saturn," Les Geisman spoke over the wind. "We've seen the Saturn 1 and the Saturn-1B, but that's only what everybody is calling the Super Jupiter and nothing compared to what we've seen on paper. This Saturn-5 of von Braun's is beyond anything ever built."

Bill Hinkle nodded his agreement but then turned to the Cowboy. "Darrell," he said. "The Michoud plant in New Orleans makes the Detroit plant at 16-mile look like nothing. In Louisiana, there are 48 acres under a roof. That's almost two million square feet."

"I flew down from Huntsville with Debus and von Braun," Hammond chimed in excitedly. "When we had that first look at that giant hanger and the Boss got excited, I realized that anything that von Braun wants, President Johnson is going to give him."

"That's for sure," Hinkle interrupted with obvious enthusiasm. "The plans for the Saturn-5 are so big the Michoud plant outside New Orleans is the only place that made von Braun happy. It's all been reworked since it closed down after making the Chrysler tank engines, and now we've got three of the big Saturn's first stages all under construction. You should see the scaffolding and the polished floors—and the lighting—the lights are everywhere—you can't see a shadow anywhere. And clean—you could eat off the floor at any part of that plant—it's that clean. There's never been anything like it."

Darrell and Les Geisman stood quietly as they listened. The thunderstorm was building but the rain was still out over the water. As a group, the four men turned away from the ocean and began walking toward the largest multi-story hanger on the planet.

With the construction of the vertical assembly building complete, the spectacular dimensions of the new Saturn-5 were beginning to take shape. When compared to the previous boosters, the size, lifting capacity, and measurements of the moon ship were startling.

The Redstone with Alan Shepard's Mercury capsule was 83 feet tall; John Glenn's Atlas with his Friendship-7 rose to 96 feet above the pad, and even the most recent Titan and Gemini combination only reached 137 feet as it stood beside the orange erector tower.

With the enormous building rising over the flat Florida landscape, all of the longtime rocketry veterans were inspired. Von Braun's moon ship booster was already achieving legendary status and the first prototype wasn't even completed. The Saturn-5 launch vehicle with the Apollo spacecraft mounted on top would rise to 363 feet above the steel-reinforced concrete launching pad.

One of the early Saturn designs, or the Saturn-1B, was eight of the already proven Chrysler Redstone rockets bundled together around a Jupiter core and strapped together with belly bands, but von Braun's moon ship was another matter altogether.

After listening to Bill Hinkle and Will Hammond, Darrell realized the Saturn-5 was basically the giant A-11 rocket that von Braun had described in the moonlight out at White Sands. The Boss might have a new government but he still wanted to explore the planets and he wanted to send a man to the moon. It was really all he ever wanted.

After another spike of lightning and a following rumble of thunder, Will Hammond picked up the pace toward the new building with the prominent NASA insignia. Everyone began walking.

"I know they engineered the new building to be hurricane proof." As Hammond spoke against the wind, he chanced a sidelong glance over to the Cowboy. "I know the idea is to keep von Braun's new baby and the moon ship safe, but did they design it to be blast proof in case one of those monster boosters blows up on the pad?"

"You bet," Darrell voiced his agreement and Bill Hinkle lost his stride. "As always," the Cowboy continued with perfected nonchalance, "there's a lot of calculation for everything. Why do you think we built the largest launching complex ever three miles away from the assembly building? We don't want to lose our new building just because we lost a Saturn on the pad."

Chapter 82

After his first year in training as an astronaut, Gus Grissom
tallied up the days that had kept him away from his home and
his family. His average day was 14 hours of training, traveling,
or public relations, and he had been away during the first year of
the program for 305 of the 365 days. He never complained about
the workload, but he was a stickler for quality engineering and
design.

January 27, 1967 - Cape Kennedy

When the first Apollo spacecraft arrived, Gus Grissom
wasn't happy. He remarked to everyone who would listen
that the new Apollo capsule was flawed and there were serious problems that needed to be fixed.

"It's not my beautiful Gemini," Grissom complained,
but he was told to stop the open criticism. When he hung a
lemon on the new Apollo capsule with the entire press corps
watching, NASA wasn't happy.

Despite tension in the trench, and with concern from several of the astronauts, the first Apollo spacecraft was mounted atop a new Saturn-1B booster and pronounced ready for
simulated flight testing. The baby Saturn beneath the new
Apollo ship was not fueled, as it sat nestled into the scaffolding at launch complex #34, because of the many preflight tests that would require all three astronauts ready at
their stations and sealed inside the capsule. In the distance,
and down the beach, Pad-39-A was nearing completion and
waiting for the first full-sized Saturn-5.

Of the original seven Mercury astronauts, Gus Grissom
was chosen as the first Apollo mission commander. Gemini
spacewalking veteran Ed White was selected as senior-pilot,

and first time astronaut Roger Chaffee was elected to fly as pilot.

Roger Chaffee was excited and thrilled to be included in the mission, Ed White was cautiously optimistic, but Gus Grissom had his doubts. The three pilots were selected to fly the first Apollo mission and pave the way for a new age of spacecraft and the Lunar Orbital Rendezvous mission to the moon. JFK's deadline to land a man on the lunar surface and bring him back safely was less than three years away.

When an overcast winter sun rose over the gray waves of a rough Atlantic Ocean, the traffic leading onto the Cape was heavier than usual. There was a full-sequence "plugs out" test scheduled for the new Apollo spacecraft, and the simulated flight analysis required that all the operations and control room crews were fully staffed. It was the beginning of the weekend but no one expected to get off early.

Darrell Loan, Les Geisman, and Les Myers all arrived at the same time and walked into the blockhouse together. It was just before 7:00 a.m. on Friday morning. The weather was cool after a cold front and everyone wore jackets against the salty winter wind. Standing next to the ocean and embraced by scaffolding and the orange gantry tower was the Saturn 1-B booster with the waiting Apollo capsule. Surrounding the new capsule and protecting the spacecraft from the elements and the ocean spray was the White Room. Well below the gantry tower and across the safety field was the entrance to blockhouse #34.

"Have you heard any more about the solar flare?" Les Geisman asked. The big Texan was shedding his windbreaker and already shaking out a Chesterfield to start the day. Between the rows of consoles and flickering monitors, the engineers and technicians at blockhouse #34 were arriving for work. The coffeemakers were already working overtime.

After Geisman's question, the control room superintendent paused before he answered. Myers was under a lot of pressure as was everyone at the Cape and at Houston. A giant explosion from the surface of the sun two weeks earlier had raised many doubts as to whether men in a spacecraft could survive such a dramatic infusion of sudden radiation and cosmic rays. Once again the critics were questioning the human factor and the very idea of astronauts in space.

"You can worry yourself to death," Myers finally responded after taking the moment to think. "But it just doesn't do any good and it doesn't help the program."

Without warning, and during the second week of the New Year, the largest and most violent solar flare ever recorded exploded from the surface of the sun. Satellites in orbit staggered under the dramatic rise in solar wind and scientists immediately began to question the safety of space travel. To Les Myers, it was just another worry, although a real concern if another solar storm happened while astronauts were in space. The walls of a spacecraft could not shield anything living from that amount of cosmic radiation.

After everyone settled into the trench it soon became apparent that the Friday morning routine wasn't routine at all. The problems over at operations and in the blockhouse started early. There were little glitches all across the board, and the slow starts to normal procedures were filled with delays and frustrating false starts. As the endless systems check slowly continued, Darrell could not stop thinking about the lopsided grin Gus Grissom gave to the press corps when he hung the lemon on the first Apollo capsule. Gus always had a mischievous and ever-ready smile, but on the day with the lemon, the expression had been troubled and guarded.

After a quick break for lunch, there were more concerns with communications, false starts, and delays in the systems checks. Grissom, White, and the eager-to-please Roger Chaffee had been dressed in the suit room and were geared up in their spacesuits when the North American Rockwell engineers held open the hatch to the new Apollo capsule and helped the men inside. The time was 1:00 p.m. and the White Room at the top of the rocket was fully staffed with technicians in white caps, white shoe covers, and spotless white coveralls.

When it was time to close the hatch on Apollo-1, Guenter Wendt was not present. The bigger-than-life German who had always been ready to tuck in the astronauts for the Mercury and Gemini missions was not working on Apollo because North American Rockwell would not give him the control he demanded. The White Room was not the same without Guenter Wendt. John Glenn had fondly nicknamed the fastidious German the "Pad Fuehrer" because of his strict demeanor and his meticulous attention to even the smallest detail. On earlier missions before an upcoming launch, when the roving media and the searching press corps would hound the awaiting astronauts to the point of harassment, Guenter Wendt would hide them at his house.

As the winter wind on the Cape churned the ocean into breaking waves and whitecaps, the protective panels that were the floor and the walls of the White Room were all in place. For anyone new to the White Room, it was hard to imagine that the weather-protected spacecraft was high above the beach and coupled to a Saturn 1-B booster. Inside the protection of the White Room, the only part of the rocket and spacecraft that were visible was the conical shape of the capsule, the open hatch, and the vast array of flight switches, instrument lights, and gauges that were clustered above and around the tightly fitted survival couches. The crashing waves on the beach were 15 stories below, but the White Room from inside looked like the sterile entrance to a simulator environment. The White Room in reality was a protected and dust-free chamber at the end of a walkway and a gantry arm that was part of the scaffolding that surrounded the nosecone of a rocket.

The open-air catwalk from the gantry elevator to the White Room was a cumbersome walk for an astronaut in a spacesuit, but the full-length view of the rocket standing beside the open ocean was worth every minute, hour, and year of education it took to get there.

After the three astronauts were in place and the hatch to the first Apollo mission was locked and sealed, everyone in the trench heard Gus Grissom's voice over the intercom speakers. He had just hooked his spacesuit up to the internal life support and oxygen system in the new North American capsule.

"There's a real sour smell in the spacesuit loop," he reported through building static. "It's a strange odor. It doesn't smell right."

After the comment, a sample was taken of the air loop and after much discussion and traffic over the intercom, everyone agreed to continue the test despite the strange smell.

As the intricate systems checks continued, the interior of the Apollo capsule slowly pressurized to an atmospheric pressure of 16 pounds per square inch of almost pure oxygen. Outside the capsule and at sea level, the atmospheric pressure was hovering at around 14 pounds.

After the fourth hour of tedious tests, the communications link became even more garbled and static filled. Darrell shook his head. "This communication system is terrible and that's with our pilots still on the ground. We really need to work on this radio link."

Les Geisman grunted his agreement and then spoke into the phone by his monitor, "I'm showing high oxygen flow in the capsule." Les was speaking to the environmental engineers.

Suddenly it seemed everyone was on the phone to someone. Les Myers was speaking into his handset, Geisman frowning as he spoke into his, and all across the control room a master alarm in the spacecraft was showing an urgent red light blinking on many of the monitors. All the blinking lights were indicating the same oxygen flow master alarm.

When Geisman hung up the phone, he looked disgusted. "The guys in environment are trying to tell me it's because they're moving; they say the master oxygen alarm is being triggered because the pilots are moving around too much. That's crazy," the big Texan argued. "They're strapped in. They can hardly move at all. They don't have any room to move."

The blockhouse speakers crackled again but this time the static was much worse than with any Gemini mission that was falling into LOS.

"What the hell is wrong?" Les Myers shook his head and picked up the phone again. "We can't hear Grissom at all, and this is in the blockhouse!" Disgustingly he replaced the receiver.

"Something's gone haywire. This is terrible. Maybe Gus was right. Maybe this thing is a lemon." Les Myers was standing while everyone else was seated at their stations in the trench. "Everyone across the board is having trouble," he offered to the room. "They're all having trouble hearing Gus and the others too. They can hardly hear anything over at operations *and* at the checkout building. This is beyond belief."

After another set of sequence starts and systems tests that were going too slowly, Darrell chanced a sidelong glance at Les Geisman. The big Texan didn't look good. He looked worried—even more so than with Gemini-6. Nothing felt right and everyone could feel it. The three astronauts had been locked in the capsule doing a mock countdown for over five hours and nothing was going by the book. If this was how the new spacecraft was going to work, the moon missions were in serious trouble.

At 5:40 the static on the com-link was so bad operations called for a hold on the countdown test.

At 6:15 the sunset was complete and the lights on the scaffolding and the gantry tower lit up the night. There was a new white rocket on the pad, a new capsule on the new rocket, and in the darkness on the beach, new waves were breaking and rushing up with a changing tide. Above the gantry tower, the overcast skies had cleared and a few stars were out, but there was no moon.

Just after the countdown hold, all but three of the White Room crew were sent down to the cafeteria and were off the scaffolding for a dinner break. The white-clad technicians were filing back into the gantry elevator when Gus Grissom's voice suddenly prevailed above the static and was understandable over the loudspeakers. "Jesus Christ," he said disgustingly. "How are we going to get to the moon if we can't even communicate between three buildings?"

At 6:40 p.m. the communications suddenly improved. The three voices in the new capsule were still scratchy, but now they were reasonably clear. As the voices became recognizable, there was a sudden surge of oxygen in all three of the spacesuits.

When the most intelligible communication of the day sounded over the speakers, everyone in the blockhouse froze. The voice was from Roger Chaffee and he sounded oddly casual.

"I smell fire," he reported over the intercom. "Fire, I smell fire."

Two heartbeats later Ed White called out urgently, "We have fire in the cockpit! We have a fire in here! Get us out!"

When the White Room technicians heard the words "fire in the cockpit," they instantly rushed to open the Apollo-1 hatch but as the seconds passed and the tension began to build so did the mounting temperature and pressure inside the capsule. Atop the Saturn 1-B booster and behind the locked hatch of Apollo-1, an electric spark in a pure oxygen environment had started the unstoppable. Before the frantic technicians could even reach the sealed and locked hatch, the spacecraft ruptured with a horrific burst of smoke and bright blue flames.

When it was appallingly clear that the three men in the capsule were in a manmade hell and beyond help, half of the soot-blackened technicians suddenly began the shutdown procedures to disarm the escape tower rocket system and the volatile solid rocket boosters. After a wall of flame divided the coughing White Room crew and the doomed Apollo-1, the horror-struck technicians battled the superheated flames with every fire extinguisher available.

The almost pure oxygen that was so flammable and so explosive was still flowing into the capsule and resembled as it vented outward through the exploded rupture as what one shaken crewmember described as the lighting of an acetylene torch.

When it became obvious the superheated flames and a torrent of gushing smoke had driven out the crew of the White Room, everyone from the trench and the blockhouse rose from their stations and ran out into the Florida night.

Near the top of the scaffolding and at the peak of the rocket, the orange flames of the fire dominated the scene until the super-heated inferno became a column of smoke that blocked out the stars.

As everyone from the blockhouse and the operations buildings ran across the safety field, it was at once clear that the voices that were suddenly recognizable only a few minutes before could now be only a memory. As the horrible realization became more and more undeniable, crews from all over the Cape began crowding into the gantry elevator. It was clear that everyone wanted to help and it was also obvious the overwhelmed, coughing, and soot-stained White Room crew needed relief.

When Darrell, Les Geisman, and Les Myers, along with the flight surgeon and many others from the blockhouse gained the elevator deck, they were crowding the lifting platform to beyond capacity. When the men from the trench finally gained access to the White Room that was now a smoke-stained disaster instead of a protective staging area atop a rocket, the opened panels and the ocean air could not take away the smell of the super-heated and burned plastic, the melted wiring, and the unmistakable scent of destroyed human flesh.

There were three brave men that had been consumed almost instantly, but the horror of the their last seconds alive — inside what now looked like a blackened furnace — was the only terror that anyone could imagine.

When the astronaut's remains had been recovered and the now blackened White Room and the capsule sealed until further inspection, the time was almost 3:00 a.m.

After everyone knew all that could be done was done, Darrell drove Les Geisman home before he turned the car toward Audrey and the kids. The big Texan didn't want to drive, and the Cowboy needed something to do. He didn't want to think about what happened above the beach on the top of the new rocket and the new spacecraft that Gus Grissom never liked.

Chapter 83

"The conquest of space is worth the risk of life."

Gus Grissom

When Les pulled back on the control yoke, the Beechcraft Twin Bonanza easily lifted free of the runway and began to climb. Even before Geisman began to turn and merge into the Titusville air traffic pattern, the view from the rising airplane was, as always, inspiring.

Just after takeoff, the endless whitecaps and blue-gray waves of the Atlantic came into view, and then the row of rocket launching pads and gantry towers perched along the beaches at Cape Kennedy.

Without comment, Les gently banked the airplane away from the tallest service tower and the new launch complex #39 that was still under construction. Also easily in view was the still-standing Saturn 1-B and the surrounding scaffolding and gantry tower of launch complex #34. There were no smoke stains or apparent damage, but it was the tragedy site of the terrible oxygen fire that killed Gus Grissom, Ed White, and Roger Chaffee.

As the Beechcraft's twin engines began to protest with opposing vibration, Geisman adjusted the throttles. The propellers then fell into sync, and the motors began to drone in oscillating harmony. The sound was somehow passive, soothing, and therapeutic. After another climbing turn, the Cape fell away and the Gulf of Mexico slowly appeared and was visible in the hazy distance.

"We'll fly up and over the coast," Geisman offered. "The plane's in perfect shape and there'll be no problems, but I always like to keep my options open."

Darrell shook his head and spoke over the droning engines, "I can't forget the smell of that smoke. I'll never forget that smell as long as I live. I've smelled it before in Korea, but this smell was worse."

"You didn't tell Audrey anything about that, I hope." Les looked over with concern to his passenger in the co-pilot seat.

"No, no, nothing about that. I did tell her that we all went up in the gantry tower to see if we could help. I explained that everybody went and we were four deep in the elevator."

Darrell looked down as the first of the Gulf beaches came into view. Les had the Beechcraft leveled off at 10,000 feet and the weather for flying was cold, clear, and perfect. The temperature outside the cockpit was just above freezing and there was a definite chill inside the airplane. The warm winter morning on the Florida coast had vanished with the high altitude.

"I told her that it was too late for anybody to help. I told her that the fire was fast. That they never knew what happened."

Geisman shook his head. "Brother, I only wish that were the truth. I'll never forget the sound of those screams over the intercom. It was fast, but not fast enough. The communication link was terrible all day but I sure heard those screams plain enough."

As the Beechcraft droned on and flew toward Texas, the final minutes of the last day at the Cape came back again and again. There were only seconds between the words "fire in the cockpit" and when the screams began. It was less than a minute until everyone came to the horrible realization that in the pure oxygen environment all three astronauts had burned to death before they could even begin to open the hatch on the first Apollo capsule.

Two of the White Room crew had to be hospitalized for burns and smoke inhalation, and many of the others from the gantry tower were still in some kind of post-traumatic shock. Before the White Room crew could open the hatch, the capsule ruptured from inside with the force of an explosion. It was the worst disaster the space program could imagine. The American people and the world could probably understand a fatal accident with astronauts atop a rocket flying at supersonic speeds, but a disaster during a ground test was almost unforgivable. There were politicians that wanted to end the space program.

When the horrible midnight of Apollo-1 was finally finished, it was well into the early hours of the morning when Les Myers had made the announcement: "Everyone should take a week off and find a private way to regroup."

Audrey agreed, and when Geisman suggested the trip out to Texas to visit his parents, she had practically begged the big Texan to take her husband along.

"Go with him Darrell," Audrey said. "Get out of here and take a break. I can't stand to see you so hurt. There's nothing either one of you could have done, and you both need to get away. You need to get away . . . to figure that out."

Audrey had made her position clear when she placed her hands on her hips and stared down at the two engineers. "You two take a break, you go out to Texas and come back here and get back to work. It's the right thing to do and you both know it!"

When the Florida Panhandle was passed and Mobile Bay another high altitude memory, Geisman spoke into the radio headset and hailed the airport at New Orleans. When he was finished with the tower and received runway clearance and instruction, he banked the Beechcraft over and began the descent.

"I didn't tell you before" the big Texan confessed, "but we're going to make a little stop and see what those boys are building out at that Michoud plant. I think we both deserve a little look at what a moon-ship booster is going to look like, and I want to see what 48 acres looks like under a roof."

Chapter 84

"After the Apollo-1 crew was lost we said that we wore a black armband for a few weeks, but we wear it in our hearts forever."

Wally Schirra

February 2, 1967 - Miles, Texas

"Happy Ground Hog Day," Les Geisman's mother announced with her German accent. "Ja, ja," she spoke with her inflection. "We are Americans now and we celebrate all the American holidays—even if some of them, perhaps, are a little strange."

"That's right mama," Les said in English, "and now I'm going to drive Darrell into town and show him around—just for the American Ground Hog Day. We'll be back before it gets dark."

With Geisman's mother waving goodbye from the massive Texas farmhouse and papa Geisman waving from a barn that easily held all of the Geisman's John Deere tractors, Les and Darrell climbed into the big Texan's latest Pontiac Grand Prix. The convertible had been stored in the barn alongside the farm equipment, and by the time Les was waving goodbye and eyeing the dust trail in the rearview mirror, he was a man on a mission.

"We're going into town Darrell to do a little drinking."

Darrell laughed for the first time since the fire, "Whatever you say Les."

As the Pontiac plowed a huge dust cloud over the West Texas plains, Geisman turned up the radio just as the announcer was saying: "*That's right folks, and next on our play list is number 12 on the Billboard country charts and Skeeter Davis singing 'Fuel to the Flame.'*"

Before either of the engineers could comment, Darrell reached over and turned off the radio. For a moment the only sound was the Grand Prix roaring through the West Texas morning. Les Geisman liked to drive fast and it was obvious he loved the Pontiac.

"I think I know who will be the first man to walk on the moon." Darrell spoke over the road noise.

"And how," the big Texan tapped his fingers on the steering wheel, "could you possibly know that?"

"Because of what happened on Gemini-8." Darrell's response was determined. It was time to carry on. He had been through the horrors of Korea and seen what happened in war to soldiers that could not let go.

"You might have a point there," Geisman nodded and the top of his head touched the canvas roof of the Pontiac. "Can you imagine being out of control and spinning through one revolution per second, ready to pass out, and still programming in a perfect reentry? That's what I call flying by the seat of your pants and keeping cool under pressure."

"That's it," Darrell agreed. "Neil Armstrong had a real emergency in space on Gemini-8 and he handled it like a walk in the park." Suddenly Darrell could smell the smoke smell again and he thought about the fire and about Korea. He thought about the soldiers he had seen in the past and how no one could really know how a person would react until there was a real emergency. Neil Armstrong, Darrell knew, had flown fighters in Korea, and anyone who had flown jet fighters over that battlefield was no stranger to an emergency.

As the Pontiac continued to cruise, Darrell pondered the flat and endless West Texas plains and the empty road. "How long Les?" the Cowboy asked over the road noise. "How long will we be delayed? Can we still make the JFK deadline?"

"We'll make it," Geisman looked over and his head touched the roof again. "All the wiring will have to be redesigned so there can be no sparks. The pure oxygen in the cabin will have to be replaced with a less volatile mixture, and the hatch will have to be completely redesigned so if there is a problem, an emergency egress will not be any trouble."

"Don't forget about the radio link," Darrell offered, "that has to go to a VHF band and a high gain antenna. If we go to VHF and high gain we can broadcast from the spacecraft with low wattage. Because it's line of sight it will work."

"It won't work when they cross behind the dark side." Geisman frowned. "That will be a total LOS situation. No radio signals can go through the moon. That's going to be spooky," the big Texan declared. "What if they go around the dark side and don't come back out. What if they disappear?"

"No radio link will work when they're on the dark side," Darrell agreed. "When they're on the far side they will be totally alone. But don't forget," the Cowboy added, "the Russians have made lunar orbit with Sputnik probes and circled around the moon since '59. If something were going to happen it would have happened to those unmanned missions."

"By the way Les?" Darrell suddenly realized the Grand Prix had been cruising through the plains and churning up dust for the last half hour. "Where are we going?"

"For drinking," Les explained patiently, "we have to go a little bit further down the road. I don't want my parents to find out we've been drinking in my hometown. Don't worry, the next town is only about another hundred miles."

Chapter 85

When the moon is in the seventh house and Jupiter aligns with Mars . . . Then peace will guide the planets and love will steer the stars.

— *From the musical Hair, 1967*

October 19, 1967

Les Geisman was in a rampage and everyone looked up as the big bear of a man burst into the blockhouse. "The Russians have landed on Venus," he announced loudly with newspaper in hand. "I can't believe it, but it's true. It's all over the headlines."

After his broadside and entrance, Geisman placed the morning paper down among coffee cups and ashtrays and stood back to watch the reactions. All the technicians in the control room were suddenly spellbound. Everyone knew there had been Soviet missions to earth's nearest sister planet, the planet that had always been mysterious because it was always covered in clouds, but the fact that the Communists had actually made a landing caught everyone by surprise.

Les Myers scratched his head. "They couldn't even get two separate ships to rendezvous in earth orbit but they can build a spacecraft that can fly all the way to another planet and land on Venus?"

Darrell was shocked. He was also impressed and suddenly curious. "What do they want with Venus?" he said. The Cowboy then picked up the paper and began scanning the print. Several of the other engineers began crowding around, including the youthful but painfully thin astronomer with the black horn-rim glasses and the pocket protector full of mechanical pencils. When Les Myers eased into the crowd, the boyish astronomer disappeared.

"It's all there in the story—everything." Les Geisman stabbed at the newsprint and then stood back to cross his arms. The big Texan was thrown and hard to interpret. It was obvious he was excited but unclear whether he was upset, impressed, or just annoyed that the Russians had made such advancement.

The bold print banner on the newspaper read: **Soviets Land on Venus. Russian Probe Sends Back Data.**

Beneath the headlines and on the second and third pages of the article the details were vague, but the information the Soviets chose to share was impressive.

Darrell handed over the paper, and the flight surgeon for the day began to scan the text. "I'll give a running commentary," the doctor offered. There were too many technicians crowding around for any single person to read the story.

"The Russians say it took 183 days to get to Venus and when the *unmanned* probe arrived they found the atmosphere was mostly carbon dioxide—99 percent." The surgeon continued to paraphrase. "They claim it rains sulfuric acid and the atmospheric pressure is 22 times that on earth. They also say the temperature on the landing surface is hot enough to melt lead. Just imagine," the flight surgeon held the paper to his chest and began to speculate, "if you happen to be a Russian in a spacesuit standing on the surface of Venus you would be squeezed to death by an atmospheric pressure equal to a couple of tons per square inch. It would be like trying to survive in a deep diving suit thousands of feet below the ocean but in an environment hot enough to cook anything in the suit."

"That's not a viable mission." Les Geisman shook his head. "Why would they even bother?" the big bear scowled. "That's an environment that could pressure-cook anything—spacesuit or not. With that description of melting lead and brimstone rain, it sounds like the clouds over Venus could actually be covering Hell."

The flight surgeon was a speed reader and after Geisman's comments, the doctor was already deep into the article and in the center of the NASA technicians. Under the florescent lights but now in the second tier among the gathered engineers, no one but Darrell saw the big Texan shiver at the thought of Venus actually being Hell.

From across the control room the youthful astronomer was back and moving through the consoles with a rolled star chart. He was determined, and parting the early morning crew. "I know why they picked Venus," the familiar figure offered as he unrolled a map of the solar system. "They picked Venus because it's so close. When Venus is the farthest away, it's 162 million miles distant, but when it's closest to the earth it's only 23 million miles. That is a 139 million mile difference. That's got to be why. The Russians chose Venus because it was so close."

"How far is Mars?" Les Myers tapped the red planet on the chart.

"36 million at the nearest point," the star navigator answered at once. "Venus is 13 million miles closer." The young man's doctorate degree was in astronomy, but in his rolled-up white shirtsleeves, black tie, and horn-rim glasses, he looked like a first year undergrad.

"They say they landed," the flight surgeon continued dryly as if the young astronomer's report was an intrusion, "but from what I can gather, I think that Les might be right and the pressure and the heat cooked their landing probe before it could send back photos or any other surface information. I'm not an engineer," the doctor confessed, "but the Russian approach does seem impressive. They used parachutes to slow their craft through the clouds of carbon dioxide and they even had the landing base rigged to float in case they landed on water."

"What's the temperature of the moon going to be like? How hot will that be?" Darrell asked the question and suddenly everyone focused on the doctor that was easily old enough to be the young star gazer's grandfather. All of the flight surgeons were well trained for the upcoming challenges and everyone knew that the medical people had a special understanding of what they were up against as far as keeping humans alive on the lunar surface.

"The moon is going to be nothing like Venus because the moon has no atmosphere," the doctor said as he rubbed his chin. "The moon also has days and nights that last for 13 of our days. That means the moon will have scorching sunlight on the same region for 13 days. With the full sun on the surface, the sector we are planning to land on—the Sea of Tranquility—will have a surface temperature of about 225 degrees. When the lunar day cycles and darkness finally turns into nightfall, the temperature can drop a couple of hundred degrees in a matter of minutes. Nightfall on the moon can get really cold because there is no atmosphere to hold in the heat."

"How cold are we talking?" Les Geisman shook out a chesterfield and everyone followed suit. This was something the boys in the trench hadn't heard and everyone was interested. Coupled with the news from the Russian Venus probe, the morning at the Cape was once again taking a turn toward the limits of the imagination. There was never a mundane or ordinary morning at the Cape. There was always more to learn, more to share, and more unanswered questions with every new discovery.

"The surface of the moon on a night of lunar darkness," the doctor continued, "can dip down to 300 degrees below zero. Our facts are fairly solid, and of course the heat will be our problem because we are landing in a period of lunar daylight, but as far as keeping our guys alive as they walk the moon, the spacesuits for the moon are being made by Playtex, the same company that makes women's bras, girdles, and underwear."

"You're kidding?" Les Myers appeared shocked.

"Nope," the flight surgeon grinned. "Can anyone think of another company better suited, equipped, or more experienced in holding important things together and protecting what's inside?"

Chapter 86

"We drill! We drill the hell out of everything. We drill and we drill until we know every step in our sleep — and then we drill some more!"

— Guenter Wendt

When the first Saturn-5 came to the Cape it came in the three sections. It was the biggest rocket ever built in America, and it was too big to arrive otherwise.

The first stage came from the Michoud plant outside New Orleans and was shipped down by barge through the Gulf of Mexico and up to the Cape via the Atlantic Ocean. The second stage came in from Seal Beach, California, and was transported by ship all the way through the Panama Canal. The third and final stage also came from California, but the Huntington Beach contribution arrived via the "Pregnant Guppy" aerospace-liner. The S4-B or the third stage of the Saturn-5 filled to capacity tested the lifting ability of one of the largest air-cargo planes in the world.

When assembled at the Cape and erected in the vertical assembly building, the big Saturn's three stages complete with the housing for the lunar-landing vehicle and the Apollo space craft rose to 363 feet and was the most complex and colossal engineering project ever assembled.

Even though the Saturn-5 was complete, with every passing day, it seemed the redesign teams for the Apollo spacecraft and the Grumman engineers for the lunar excursion module wanted to increase their payload weight further. The new innovations made both of the spacecraft better and safer, but the last minute changes also added additional weight.

Everyone now knew that von Braun was right and it would take a massive three-stage rocket to lift the payload that was necessary to make it to the moon. The weight and payload numbers were still increasing, but the German's earliest estimate of a fifty-ton lunar payload appeared as a perfect forecast. His life had been dedicated to the moon ship, and his earliest weight and redline figures were as always right down the middle. He had already considered the spacecraft designers and excursion engineers would want to make last-minute additions, and he had compensated with a booster that could lift 125 tons into a low earth orbit.

The Russians, of course, were working franticly to outpace the U.S. efforts to the moon and also had a manned lunar program and massive three-stage rocket. The Russians were calling their rocket the N-1, and their spacecraft "Zond," but the Soviet model—according to the latest Smith and Jones intelligence reports—could not manage to get off the ground. Apparently, there had been more than one launching pad explosion and the pressurized fuel detonations had been beyond horrific. The Russians, however, were not the only ones having explosions.

"I talked to Will Hammond about the explosion in Huntsville," Darrell said as he walked with Les Geisman toward the vertical assembly building. The Florida summer was over and the October weather a welcome relief, but the Cowboy could not stop thinking about the Russian disasters and the latest worries from Huntsville. "Will said the explosion blew the test stand to pieces, and they're still finding parts from the S4-B all over the place."

The big Texan nodded as he walked, and it was obvious he was deep in thought.

A full throttle engine test of the new Saturn-5's third stage—or the S4-B—had been almost as disastrous as what happened in Russia. The third stage explosion in Alabama had been terrible, and Russian technicians had been killed in the launching pad explosions in Kazakhstan. Lifting 50 tons to the moon was pushing the limits of all known engineering and everyone on both sides of the globe was feeling the pressure. Americans had been killed in Apollo-1 and now Russians at their launching facilities.

After the long walk in the autumn sunshine and time to contemplate the latest setback in Huntsville, the two friends from the trench arrived and were standing inside the vertical assembly building and looking up at the first Saturn-5 ever assembled. The vertical access door was open to the elements and the October afternoon was producing the very best of low humidity and perfect Florida sunshine. Inside the assembly building, steel beams climbed to meet struts and rafters as the cavernous structure rose all the way to five hundred feet. Along every tier of the rising and massive workshop, rows of spotlights were focused and shining on the first moon ship in its hanger.

Geisman was silent as he looked up at the rocket. There had never been anything even remotely close to what was standing in the hanger with the support cables of the overhead cranes.

"I'm not so sure about George Mueller's idea," Les offered quietly. "No one has ever done this before and insisted on a full 'all up' test. Just look," the Texan tilted his head, "it's obvious that the Boss is not happy."

Across the polished concrete floor of the assembly building, von Braun was standing with Kurt Debus and a knot of anxious engineers surrounded the two Germans. The Germans were, as usual, dressed in their dark blue suits with white lab coats and the technicians in a swarm of blue and white coveralls. Von Braun was shaking his head as he glanced at an offered clipboard. After a moment, Kurt Debus turned away from the debate and went to stand under one of the massive F-1 engines on the Saturn-5's first stage.

There were five of the F-1 Rocketdyne engines, and each one of the largest rocket engines ever built weighed ten tons. The base of the engine bell on each exhaust cone was 14 feet across and the apex of the cooling jacket and gimbaled engine head was over 20 feet tall. Together, the five F-1 engines would burn 15 tons of liquid oxygen and kerosene per second. The engine bell of just one of the F-1 engines could cover the entire group of technicians surrounding Von Braun.

"What did Hammond say caused the Huntsville explosion?" Geisman asked the question but he seemed fascinated with Debus. The German was just standing quietly and looking up at the five colossal engine bells clustered together. He was apart from his engineering group and just standing and staring upward into the gimbaled engine cluster.

"It was a faulty weld in a spherical helium tank," Darrell answered, but he was also watching Debus. "He said the fuel detonation explosion was heard for miles around and the local townsfolk are nervous. It was the worst failure the Huntsville people ever had."

"That makes me think again about the 'all up' test." Geisman shook his head and stuffed his hands into his pockets. "If we can't get just one stage to operate without trouble, how can we expect the whole Saturn-5 with over five million parts to fly without a glitch? If something goes wrong, how will we know what happened or where?"

"You know telemetry can tell us all that." It was Darrell's turn to shake his head. "I like the idea," he said. "We might as well find out if this monster is going to work, and we might as well see if she's going to fly with all three stages stacked together. Time is running out for the JKF deadline."

During the entire ballistic missile program and all across the drafting boards of all the engineers, the theory had been "Baby Steps." Take small steps forward and check everything before it was time to take the next step. Baby steps in rocketry meant that with every redesign there needed to be a new test even if the new change was small.

George Mueller was one of the top administrators at NASA and his "all up" idea, or "test all the parts at once," was a drastic and unheard-of concept that would save time and money. The idea was new and radical and something von Braun and the other Germans did not favor. But as the time and money constraints were coming under more and more serious scrutiny and the political pressure was building, "all up" testing was now considered a reality and only two weeks away.

The first Saturn-5 ever assembled, standing in the hanger with Kurt Debus standing directly under hundreds of tons of hanging fuselage, was designated Apollo-4 and was scheduled for a full flight but unmanned test in less than ten days.

After a few more minutes with the assembly technicians, von Braun looked up and found that Kurt Debus was not at his side. He then looked over and smiled when he saw his colleague standing under the massive engine cluster. With the gaggle of engineers beginning to follow, von Braun handed the clipboard back to the chief engineer and headed for his friend that was waiting under the rocket. As a group, the technicians respectfully stopped just before the base of the big Saturn when they understood what was happening.

Von Braun joined Debus under their creation, and as the two Germans stood side by side, they looked upward and into the heart of the rocket engines. There were no words spoken, but the technicians in the coveralls knew they were not to take another step forward. It was time for a moment of reverence and everyone suddenly knew it. Les Geisman, the Cowboy, and all the engineers in the largest aircraft hangar in the world now knew that the Boss and Kurt Debus were taking a moment.

The mood was suddenly contagious and began spilling out even to the men working aloft in the scaffolding and to the others on errands in the base of the assembly building. For a few solemn moments, everyone stopped working and quietly stood with their hands together. Kurt Debus had felt it first, and the emotion suddenly became infectious. It was time to stand back and take a moment, and everyone felt it together. Standing and assembled for the first time was a pinnacle of human achievement. There was not only a rocket but an assembled dream that was about to come true. Painted perfectly on the side of the first stage of the booster were three large red letters in a vertical line. Standing out proudly on a corrugated white background, the letters from top to bottom read: USA.

As suddenly as the moment came, it was over, as von Braun clapped Kurt Debus on the shoulder. The smaller German with the jagged scar on his chin and thinner frame then nodded and looked back to the group of assembled technicians. Before von Braun turned back to the waiting men, there was a very fast movement from his right hand to his eye. From the distance, no one could be quite sure, but for a single second, it appeared as if the Boss had suddenly wiped away a few tears of overwhelming passion.

Chapter 87

Where there is no vision, the people perish.

— Proverbs 29:18

November 8, 1967

"Douglas aircraft has plans for a manned mission to Mars before 1974," Les Geisman offered.

Les and Darrell were outside the blockhouse for #39-A, and watching the latest fueling procedure. They were not alone. Many of the control room engineers and technicians were watching. The first Saturn-5 complete with the third stage housing for the lunar landing vehicle and a boilerplate Apollo spacecraft were holding court as a string of tanker trucks waited in line to fuel the largest rocket ever built. It was after just midnight and the launch pad, the gantry tower, and the moon ship were illuminated brighter than daylight with a hundred spotlights.

"The Saturn-5 can boost 125 tons into earth orbit and 50 tons into lunar orbit. Mars would not be out of the question. We could do it," Darrell paused to look aloft and ponder the stars that had never looked brighter. "In fact," the Cowboy declared, "I'm beginning to believe we could do almost anything. The Saturn-5 can boost a payload to any planet in the solar system."

Although most of the Cape and launch complex #39-A was bright with artificial light; the moon was still full and very prominent, and Mars a steady beacon sparkling over the horizon.

"Did you hear about the Russians super booster for the moon?" Geisman asked. "They say it has thirty engines, just for the first stage."

"They also say they haven't been able to get one off the ground." Darrell's response was guarded. He had been briefed again about another of the Soviet failures and he also had doubts as to whether the precise management of 30 rocket engines firing at once was possible. There was a satellite photograph going around showing an entire Soviet launching facility destroyed by an exploding super booster. The Russian version of von Braun's moon ship was still a disaster, but the Communists had already sent a smaller rocket with a lighter payload to orbit the moon and it had reportedly returned safely. The "Zond" mission for lunar orbit carried living turtles to determine if they could live through the flight, but there were conflicting reports as to whether or not the turtles had survived.

Darrell smiled as he looked aloft. There had been Communist turtles circling the moon.

"Ready to go?" Geisman tilted his head with the question. It had been a long day. The sequence checks had been relentless, and as Darrell nodded and the two friends began walking, the Cowboy stretched his shoulders and then his back.

"How about a stop at the beach for a beer?" Les looked over.

"It's after midnight," Darrell offered as the partners from the trench continued to walk.

"I know," Gesiman confessed, "but I just can't seem to sleep anymore—everything is going so fast—besides the Starlight lounge is open until two."

"Okay," Darrell conceded, "but we're just stopping for one. We're launching this monster tomorrow, and for better or worse, it will be one of the biggest days in the program."

*   *   *

The tension on the Cape during the last few weeks had been beyond intense, but somehow all of the accomplishments and the men and women that had made those accomplishments possible made everything worthwhile. It was hard to believe so much had happened after the fire—and everything really was either "before the fire" or "after the fire"—but six months after the fire and the horrible deaths of three astronauts, the unmanned Apollo-4 countdown became complete and the mighty Saturn-5 made it's dynamic debut.

As the Greek God of light and a blazing white exhaust trail that appeared even bigger than the base of the rocket would allow climbed above a bright orange fireball, the dramatic introduction became complete as the thundering sound waves stormed over the Florida peninsula and rattled windows for a hundred miles.

"Wow!" Walter Cronkite spoke to the CBS television news audience during the first moments of the first Saturn-5 launch. The news anchor was sitting at his desk with binoculars, and he and the audience had a full camera view of the rising Saturn when the tremendous battering wall of thunder abruptly invaded the CBS studio.

"Would you look at that rocket go!" the veteran newscaster reported with so much emotion his voice was trembling.

"That's just terrific ladies and gentlemen," Cronkite was bursting with national pride. "We are actually having some of the ceiling tiles fall from the roof! The vibration of that first stage and the wall of sound that has engulfed our studio is beyond believable. This is truly incredible," the CBS anchor declared as the Saturn continued to thunder. "There has never been anything . . . even remotely close . . . to what is rising above the Cape right now."

The sound and vibration waves had been so intense that dust had fallen from the ceiling of the blockhouse, and the CBS newscasters were forced to hold a vibrating plate glass window in place for fear that it would shatter. Walter Cronkite and his CBS news crew had been three miles away from launching pad #39-A when the first Saturn-5 claimed the sky.

Chapter 88

From the earth to the moon:

Astronaut Jim McDivitt (looking at the Lunar module with Grumman engineer specialist Tom Kelly): "She's a beautiful machine Tom."

Kelly replied: "Isn't she."

Astronaut Rusty Schweickart (as he and McDivitt walk away): "You really think it's beautiful?"

McDivitt: "God no. It looks like a toaster oven with legs but I'm not going to tell him that!"

October 1968

When Apollo-7 launched perfectly atop a Chrysler-made Saturn 1-B booster—which was eight of the Old Reliable Redstones bundled together to make the smaller Saturn's first stage—the first flight for the Apollo spacecraft began with a joke and once again reflected the boisterous behavior of the American astronauts.

After Guenter Wendt, back on the job as White Room leader, closed and sealed the hatch on Apollo-7, the command module pilot Donn Eisele looked over to his commander Wally Schirra and his co-pilot Walter Cunningham. The rookie astronaut with the ever-ready smile was bursting with anticipation as the three pilots waited side by side, but it was not until the countdown was complete and the rocket began to rise that he delivered his joke with an exaggerated German accent.

"I vonder," Eisele asked as the booster vibration began, "vhere Guenter vent?"

Wally Schirra laughed most of the way through the first stage burn. After staging, and with America's first team of three pilots headed into space together and riding a new liquid hydrogen fuel for the first time, Wally transmitted to his Gemini-6 partner Tom Stafford who was the Cap-com in Houston. "Tom, she's riding like a dream," Schirra reported from 70 miles high as the three astronauts were still accelerating.

Apollo-7 was on an 11-day mission to orbit the earth, and for the longer flights the control room and trench crews had been assigned into separate eight-hour shifts. There were color codes assigned to each crew and each shift.

Darrell, Les Geisman, and Les Myers were on the white crew. The other crew colors were Green, Black, and Maroon. As the race to the moon intensified, NASA became more and more determined to not make any mistakes. A rested and refreshed crew was always sharper. NASA had also decided, with a little influence, that the best player in the right position always won the day.

Guenter Wendt had returned as pad leader of the White Room because Wally Schirra had gone all the way to the top NASA brass and demanded that the fastidious German was reassigned as White Room leader for all future flights. According to the rumor mill, Schirra never officially said it— but he certainly hinted that if Guenter Wendt had been on the job—the Apollo-1 fire might have never happened.

During the flight of Apollo-7, and what everyone in the control rooms called the Walt, Wally, and Donn show, morale at the Cape soared when the Apollo service module engine fired for the first time and slammed all three pilots back in their seats. With a G-force pressure building faster and stronger than anyone anticipated, Wally Schirra once again won the hearts of the control room crews when he transmitted over the intercom with a Fred Flintstone voice. "Yaba-daba-doo," the veteran astronaut voiced during the rushing burst of speed.

When the Walt, Wally, and Donn show splashed down near Bermuda and only two miles from the planned recovery site, JFK's mission to the moon appeared to be back on track — until the first Lunar-lander arrived and miserably failed several key points of inspection. It was a discouraging discovery in comparison to the recent Apollo success, but there were some serious problems with the spider-looking spacecraft that had been designed to land on the moon.

After the news of the failed inspection, most of the engineers from the trench joined Les Myers and were standing in one of the side bays of the vertical assembly building. Everyone was looking with a critical eye at the rejected Lunar Excursion Module that had failed the preflight inspection and was *not* going to the moon.

The moon machine looked like a mechanical spider with four huge legs and two triangle-shaped eyes. There was a mouth-like hatch centered below the window ports where the astronauts would emerge and descend a ladder to the lunar surface, but somehow, the frowning opening didn't look big enough, and the machine overall appeared extremely ungainly and fragile. The fact that the landing legs and many other parts were covered in crinkled gold foil was another feature that seemed too delicate, and as Darrell, Les Geisman, and many others from the blockhouse walked around the lunar-landing-vehicle — that would *not* be loaded atop the big Saturn's third stage — it was easy to vent frustration toward the Grumman technicians that had pronounced the first Lunar Excursion Module ready for spaceflight.

"They say that in some places," Les Geisman offered, "the hull — if you can call it that — is only slightly thicker than three layers of aluminum foil folded together."

From the blockhouse crew that were now stalking and circling around the lunar spider like Indians around a Wild West wagon train, the sharp arrows of the engineers comments echoed around the storage bay and drifted up to the rafters. Adjacent to the side bay where the lunar landing craft was exiled, the latest Saturn-5 was being erected and assembled in the largest part of the hanger. Von Braun's rocket was ready but the lunar landing vehicle was not.

"I've heard that a technician dropped a screwdriver once," the youthful astronomer with the horn-rimmed glasses recalled, "and it fell right through the flight deck and landed on the hanger floor. The flight deck is that fragile."

"Now," the stargazer added solemnly, "when a technician goes inside, they have to wear big soft furry shoes so they don't kick something and make another hole."

"Just imagine," another of the crew called for attention as he walked up to the Grumman machine and tugged at the crinkled gold foil, "what if one of our boys turns wrong and pokes a hole into the vacuum of space?" The young engineer shook his head sadly.

"Their blood would boil," the flight surgeon offered sagely as he stepped up to the moon machine. "That's what would happen if they accidently poked a hole through that scotch tape and aluminum foil and made an opening that vented into the vacuum of space."

"When they first tested the LEM in a vacuum, one of the windows blew out. They still haven't figured out exactly why. And they just changed the name from 'Lunar Excursion Module' to 'Lunar Module' because the guys at Grumman figured that 'excursion' sounded too much like a fieldtrip for Boy Scouts or a grade-school outing."

Darrell was standing between Les Myers and Les Geisman but the Cowboy remained silent. He was thinking about Kurt Debus and von Braun as he had seen them earlier in the side bay looking at the failed lunar-landing craft. The Germans did not seem outwardly upset with the Grumman setback. They did, however, seem excited and impressed with the lunar spider design that eventually would ride hidden away on the third stage of their rocket.

"Well," Les Myers began as he looked around to gather his colleagues, "I can see that most of us are all on the same page, and I thought we might as well have a look at the reason Deke Slayton and the boys upstairs have changed the whole Apollo schedule. The folks at Grumman are now saying they won't have a fully operational lunar module ready for earth-orbit testing until February of next year."

When the control room superintendent's voice echoed and died away in the cavernous chamber it was followed by a moody silence. Everyone knew that the Grumman timetable was too much of a break in the program, and Kennedy's goal for the moon was suddenly unrealistic. It was as if the last nine years suddenly fell short and a thoroughbred racehorse stumbled just before the finish line.

Before Myers began again, he took another look at the 18-ton spider for the moon and stuffed his hands in his pockets. Then he nodded and continued with a newfound confidence, "What George Low, Deke, and even Mr. Webb have decided is that we are going forward with a manned mission to the moon for this December. We are going forward without the lunar landing and heading for lunar orbit with just the command and service module." Myers paused as he looked at his crew, "And we're damn sure bringing our boys back home for New Year's."

"Les," the big Texan who also shared the same first name stepped forward. "This almost sounds like the plan cooked up for Gemini-6 and 7—even with the timing at Christmas."

"It's also going to be Frank Borman and Jim Lovell again along with the new guy Bill Anders," Myers continued. "Apollo-8 is going to be the same crew for Gemini-7 but with a new guy for good measure."

After the official news drifted away and blended into the sounds of work in the assembly building, Les Myers offered softly and just for the members of his crew: "Let's hope we have the same luck we had with the Gemini Christmas for this holiday season . . . because *this* time we're going to the moon."

Chapter 89

"It is difficult to say what is impossible . . . for the dream of yesterday is the hope of today and the reality of tomorrow."

— Robert Goddard

When Darrell packed Audrey and the kids into the family Chrysler, no one knew what to expect. The Cowboy was fully immersed in his practiced and patented role of Mr. Calm and Nonchalant, but there was a casual urgency in his manner that suggested there was no time to waste. Darrell kept looking at his watch.

As the '68 New Yorker cruised up the beach from Titusville toward the Cape, Audrey searched through the local radio stations.

"That latest song was 'Credence Clear Water Revival' ladies and gentlemen," the DJ happily offered, *"and a song that has special meaning to everyone here at Cocoa Beach. Its number 24 on the Billboard charts and the title is 'Bad Moon Rising' . . ."*

Before another remark was possible from the Cocoa Beach radio show, Audrey quickly turned the dial and searched for another station.

Darrell could not help but smile at the gesture and once again glanced at his watch. The timing was perfect. He then took a quick look in the rear-view mirror to check on the kids. As usual, Deb and Mike were in the backseat and looking out to the beach as the dunes and the breaking waves appeared between groves of palm trees. The surf was breaking big on Cocoa Beach and with the windows open and the ocean breeze strong, the scent of the salty spray was tangy and unmistakable. Kath, as usual, was sitting in the front seat and straining to look over the dashboard.

On the next station playing music was The Archie's singing "Sugar, Sugar," and Audrey settled back to listen. She knew that whatever Darrell had planned was going to be good because she had learned over the years that whenever there was a quiet but determined suggestion from her husband it was going to be worthwhile.

The setting sun was all but a glow in the west as it settled lower over the peninsula, but there was still enough light to see the launching towers and the vertical assembly building. As the Cowboy offered his Crypto security clearance to the guards at the NASA security gate, the low aircraft warning lights began to blink red on the distant gantry towers. Darrell knew most of the entry guards from the early morning and daytime shifts, but the four security officers that guarded the southern entrance were new. The leading officer did however wink and smile at Kath in the front seat as the Chrysler was cleared and motioned through. There were several other cars ahead and a few behind, mostly with NASA officials and their families, and it was suddenly obvious that it was no coincidence that everyone was arriving at the same time. Everyone in the program had been through a lot and now something special was about to happen. It was time for something extraordinary.

When the car was parked with the others near the vertical assembly building, Deb and Mike were the first ones out; they knew something was coming and they could feel the excitement. There was a sense of anticipation that surrounded the Cape and it was enhanced even more by the perfect winter weather. Air-conditioning in cars was a blessing during the heat of the Florida summer, but during the holiday season there was nothing better than a salty sea breeze coming in off the Atlantic.

The mood over the crowd was clearly festive with splashes of red Coca-Cola coolers and soft drinks on ice. Blankets appeared and were placed on trunk lids for seating, and as the twilight began to fade and the darkness began to gather, there was a noticeable measure of pride and ownership as everyone settled into the early evening atmosphere that had the feel of a company picnic. This was an unofficial gathering of NASA technicians and their families, but there was a distinct feeling of achievement that was easy to recognize.

As Audrey looked around the milling families and the field of cars, and as she sat and embraced the ocean breeze and the beautiful December evening, she was determined to best Darrell at his perfected persona of nonchalance. If the Cowboy wasn't going to tell the story, spill the beans, or offer any explanation for the outing, Audrey had already decided she could wait her husband out.

"Daddy what's going to happen?" Deb asked after she was settled and content with her Coke and her seat on the Chrysler.

"Tonight we're going to see something special," Darrell offered as Audrey and the kids focused to listen. "There's going to be a lot to watch during the next few days, but tonight is special and it's just the beginning. Tonight, this is just for us — for us and everybody else that wanted to bring their families to watch."

As darkness continued to settle over the Cape and the aircraft warning lights continued to wink and blink red signals on the gantry towers, a full moon slowly began to crest and then rise beautifully out of the Atlantic Ocean. The big winter moon was a mix of orange and amber as it rose over the waves, and the path of light that spilled over the water was a shining roadway that led all the way to the Sea of Tranquility.

Just as Audrey was considering that the trip to watch the moon rise was nice and that the new-age NASA families were a tightly molded group that liked to share special moments together, she heard a low rumble and felt Darrell's arm around her as he squeezed her shoulder.

When Audrey turned to follow Darrell's gaze and look toward the low rumbling sound, she felt her heart race and goose bumps rise against the balmy evening air. She also felt her emotions soar with the unmistakable pride that suddenly flooded over all that were watching. All the children that had been running around the cars playing tag had suddenly stopped in their tracks, and as everyone turned to the giant building with the open doors, the multiple spotlights that were now shining on the 40-story rocket took all the glory from the moon rise and the face of Mother Nature.

She knew the rocket was going to be big—Darrell had told her—but what she now saw moving out of the building on the massive crawler machine was beyond anything she could have ever imagined. The moon of course was still shining and the path over the ocean was still silvery and bright, but as the illuminated rocket, gantry tower, and giant treaded crawler inched slowly forward, Audrey felt another rush of emotion and tears splashed on her cheeks.

She now knew as she quickly looked around that all the others were having the same bursting sensation of pride. The feelings she and the others now shared were such a complex mixture of passion and pride that there might not ever be another moon rise quite like this one. And yet at the same time she knew that—God willing—this was only the beginning of the greatest thing that humankind could ever dream of or that a civilization could ever accomplish.

"Oh God, thank you Darrell," Audrey sobbed and the Cowboy squeezed her shoulder again. "I just never expected to ever see anything like it. It's so beautiful and terrible all at the same time, and it makes me hurt to think how powerful and dangerous and delicate it is—all at once. This is America at its best, and I'm so proud of you and of all the thousands of others and everyone from all over the country that made this happen. I've never been this excited and scared all at the same time. Oh Darrell, will it work? Tell me it will work and not blow up like all the ones we saw on T.V."

After another soft squeeze the Cowboy offered quietly, "Of course it will work, and it's going to be the greatest thing that's ever happened. It's not just about the three men that are going to ride that monster; it's about all of us that will be riding along too. All of America is going to ride in that ship all the way to the moon. It's going to be the greatest thing since Columbus and maybe the greatest thing ever."

Chapter 90

Green, White, Black, and Maroon . . . Retro, Fido, and Guido . . .
and all the others . . . May the good Lord ride all the way.

Predawn, December 21, 1968 - Cape Kennedy

After Frank Borman, Jim Lovell, and Bill Anders spent
the night at the Cape Kennedy crew quarters, three doctors
awakened them at 2:30 a.m. for a preflight physical. When
the medial team was finished and the three astronauts
had been thoroughly prodded and probed, they were pronounced perfectly fit for spaceflight travel. It was then time
for a breakfast meeting with three other Apollo astronauts
and the top NASA brass.

Neil Armstrong and Buzz Aldrin were bright-eyed and
ready for anything as they were the backup crew for Apol-
lo-8. Astronaut Jack Schmitt was present along with George
Low and Deke Slayton. George was the Apollo program
director and "all up" testing advocate and Deke Slayton
the director of flight crew operations. Everyone had Alan
Shepard's traditional preflight breakfast of steak and eggs
with coffee, toast, and tea.

After breakfast the three pilots were meticulously suited
up in the spacesuit room and were transported by van to
pad #39-A. The three pioneers carried their portable air supply and oxygen systems as they walked — with the cumbersome hoses attached — and stepped into the gantry elevator
that would take them to 320 feet above the launching pad
and the waiting crew in the White Room. Every group of
technicians they passed paused to give a smiling thumbs-up
or a round of applause.

When the three astronauts for Apollo-8 crossed over the gantry arm above the Cape and were high above the breaking waves, the beach, and the open ocean, they were the first pilots ever to board a Saturn-5 launch vehicle. After the astronauts crossed into the White Room and the clean room door was sealed, there were a few very unofficial moments that meant more to the pilots than any speech a politician could ever give.

Guenter Wendt stood quietly for a moment alongside his crew in the white coveralls before he stepped forward and gave a bear hug to first-time astronaut Bill Anders.

"Ja, Ja," Guenter began with his strong German accent, "may the good Lord ride all the way." It was the quietly voiced prayer from Tom O'Malley when he pressed the firing button on John Glenn's Atlas booster, but the prayer and the sentiment had caught on.

"And for you two, my boys of Gemini-7," the veteran pad leader added with his accent as he also hugged Lovell and Borman.

Jim Lovell grinned and winked at Bill Anders when Borman was getting the hug.

When the touching moment was finished and a photographer documented the occasion, Frank Borman, as commander, climbed down and through the open hatch and took his traditional left seat in front of the gray instrument panel of softly glowing lights, gauges, and latch-proof and covered switches. Entering the Apollo capsule was like climbing into an aircraft cockpit from an open roof hatch. Everything was situated below. Bill Anders followed to sit on the far right after the commander was settled, and afterwards, Jim Lovell climbed in and took his seat in the remaining survival couch between the other two pilots.

As always, Guenter and his crew in the White Room wore their spotless white coveralls, white caps, and white shoe covers with more pride than it was possible to imagine. When the three astronauts were settled and fitted into the custom fitted couches, a gentle white room foot was placed on the chest of each pilot as the safety straps were pulled, tugged, and tightened to full capacity. When the pad leader and his meticulous technicians were finally finished, the flight team for Apollo-8 was thoroughly strapped in and ready to pilot the most powerful launch vehicle and the most advanced spacecraft ever built.

Just before the hatch was closed, Guenter motioned his senior technician aside and leaned into the spacecraft to look at each man carefully. All three of the pilots were already busy with checklists and beginning to select and reset the preflight switch positions that had been trained into them as rote, but the tender-hearted German, as always, needed another moment.

Without being obvious, Guenter sniffed the air in the capsule as a precaution against the unthinkable fire and then listened for a misplaced hiss of venting trouble as he slowly scanned the interior of the first flight deck that was headed for the moon.

Inside the spacecraft, the lighting was perfectly soft and restrained but without the slightest chance of a shadow possible. Throughout the supple gray and white interior of the cabin, the green illumination glowing on the instrument panel dominated the scene, but in the air above the men there was the quiet determination of heroes.

As the three pilots paused to look up to the fastidious German with the white cap and black horn-rim glasses who was watching over them, there was a sudden realization that would never be found in any training manual. Strapped into their seats and lying on their backs, the astronauts suddenly understood they were positioned in front of the most complex command console ever assembled, and they also knew they were seated before the instruments that would guide them to their destiny.

"Is everyone happy with the tension on your straps?" Guenter asked softly as he began the ritual. "Does everything smell right and feel right? Does everything sound right? Is everyone completely happy?"

Frank Borman looked first to his partner from Gemini-7 and saw the man he had spent two weeks in orbit with circling the globe. No two pilots had ever experienced anything like the cramped quarters and flight durations of

Gemini-7. Lovell just grinned and nodded his head. Borman then looked as a commander to Anders who had never been into space before.

The younger man quickly nodded, swallowed, and found his voice. "Yeah," he said, "I'm okay."

Guenter then reached in and shook hands with each man and quickly offered, "Good luck guys." Then he was gone, the hatch was closed and locked, and the countdown and the checklists continued.

Borman and Lovell were all confidence and proficiency as they continued through the preflight and countdown lists, but there was something more; something that couldn't be measured. Bill Anders' actions and responses for his role were precise, concise, and perfect, but he didn't know what was coming.

After the initial switch positions were checked, next on the list was the thruster controls that would pitch, yaw, and roll Apollo-8 while on the 240,000-mile journey to the moon. If the preflight thruster check wasn't perfect, Apollo-8 wasn't going anywhere. Being strapped into a space capsule with the hatch closed and sealed, as Borman and Lovell knew, didn't actually mean that liftoff was going to take place as scheduled. A lot could go wrong with a 6.2-million-pound rocket and Wernher von Braun, Kurt Debus, the boys in the trench, and a thousand others that were the links in the chain would not take any chances.

Outside the White Room and the sealed command module of Apollo-8, the Saturn-5 and the Apollo spacecraft were standing 40 stories above the beach beside the orange gantry tower and were ablaze in spotlights. The lighted icon of a new age was clearly visible for miles, but the thousand pounds of ice that were covering the booster stages as the pressurizing liquid oxygen and hydrogen froze the Florida humidity into a frosty glaze was only visible to Guenter Wendt, his hand-picked crew, and the other technicians that were close enough to know they were standing beside a monster.

Climbing above the waves on the Atlantic Ocean, the radiance of the false dawn was already turning into first light as the pink rose of the clear December morning was finding each of the faces that were gathered, waiting, and watching.

Over all the loudspeakers at the Cape and beyond the viewing stands where half a million spectators were anxiously crowded at the beaches, the voice of public affairs officer Paul Haney echoed the official voice from NASA. As Haney sat at his desk and spoke into the microphone, his message from the Cape was transmitted to all the radio and television stations across the Free World. All along the coast the message from NASA echoed from car radios and hand-held transistors as the thousands waited, watched, and listened carefully.

"This is Apollo Saturn launch control. We are T-minus one hour and thirty minutes and counting. The Boost Protective Cover was just placed atop the hatch on the Apollo-8 spacecraft just moments ago, and the White Room crew are securing the astronaut boarding area. Once the White Room crew does depart at a designated time, the swing arm that is now attached to the spacecraft and the White Room at its tip will be moved some three feet from the spacecraft, and it will remain in that standby position until the T-minus 5 minute mark in the countdown. The purpose here is to have the White Room standing close by in the event that the astronauts will have to be removed from the spacecraft. In the event that an emergency condition develops, which would require the astronauts to depart from the spacecraft, we could bring the White Room in from just three feet away.

The astronauts aboard the spacecraft are now participating in a test of the stabilization and control system of the Apollo-8 spacecraft. As they move their hand controllers, which would provide maneuvers in space, we are checking the performance here on the ground. All aspects of the mission are still Go, weather is satisfactory, the various tracking elements all Go at this time — T-minus one hour twenty-eight minutes twenty seconds and counting.

The countdown has been going very well since it was picked up at 10:51 Eastern Standard Time last night. Shortly before we resumed the count, the 9.8 million pound Mobil Service Structure was moved to its park position some 7,000 feet from the pad. About an hour later we began the cryogenic propellant loading of the Saurn-5 launch vehicle. In some four and a half hours we load-

ed close to a million gallons of liquid oxygen and hydrogen aboard the three stages of the Saturn-5. We now have a vehicle standing to 363 feet and weighing 6.2 million pounds on the launch pad here at the Kennedy Space Center. We are continuing to top off the liquid oxygen and liquid hydrogen supplies because they must be maintained under extremely cold temperatures. These elements will continue to boil off and we will continue to replenish these fuel supplies down to the final minutes of the count. This is Launch Control."

After the latest announcement became part of the ongoing background chatter, Les Geisman tapped a pencil and Darrell looked up when the latest thruster test was complete. Les grinned as he nodded to the Cowboy and glanced along the front row of monitors. Wernher von Braun and Kurt Debus were marching down the trench. They had never looked more intense, focused, or happier. It was the biggest day ever for the Boss and Kurt Debus, and everyone knew it.

With more coffee refills and cigarettes that anyone could count, and with the telemetry streaming in from all three stages of the rocket and the spacecraft, and with the redline values flickering perfectly on all the monitors, Paul Haney once again sounded over the speakers in the bustling control room.

"This is Saturn Launch Control at T-minus sixteen minutes and counting. The Apollo-8 space vehicle is Go for our planned liftoff at this time. This was confirmed by spacecraft commander Frank Borman."

Chapter 91

During the Apollo-8 liftoff, Commander Frank Borman reported: "First stage was smooth. This one's smoother."

Jim Lovell: "Smooth and smoother? Who are you kidding?"

Bill Anders: "Yeah, that first one felt like we'd been in a train wreck!"

Frank Borman came back with a grin: "Kick in the pants, huh?"

"Here we go," Les Geisman said under his breath. "May the good Lord ride all the way."

"Amen to that," Darrell echoed quietly and then focused on the latest azimuth release on the Saturn's internal guidance system. The big stacked rocket needed a point of reference when released from the pad and at 16 seconds before liftoff, Darrell sent the signal to the instrument unit in the booster. Now the monster had its final bearings in relation to the earth as the globe was continuously turning. After the Guidance Reference Release Event, the Cowboy could take a moment to breathe.

With the teletypes in the control room continuing to clatter, the Pan Am girls stood transfixed as the intercom speakers sounded smoothly with the very controlled voice of Paul Haney. *"T-minus 15, 14, 13, 12, 11, 10, 9, and we have ignition sequence start."*

Outside the blockhouse the tranquil beach scene exploded with a sudden flash of white-hot and then bright orange light as enormous sound waves began to roll and thunder. In less than a second, the standing Saturn-5 had completely dominated the central Florida morning. The rumbling wall of sound that followed the orange burst of light and the white hot steam felt more like a deep earthquake than anything that could be artificially created from above the ground.

Pelicans and seabirds flying on their early morning routine from the rookeries on the Banana River to the nearby beaches dove for groundcover as the first and the biggest of the three Saturn stages exploded into life.

As the blockhouse windows were suddenly filled with the orange flash, the eruption of white smoke and steam, and the beginnings of what felt like a seismic event, the calm and clear voice of Paul Haney continued through the thunder, *"The engines are on, 4 . . ."*

When the directed exhaust flow from the five engine bells engulfed the gantry tower, and as the battering thunder jumped the distance from the concrete pad to the interior of the spacecraft and as the hydraulic clamps that held the Saturn to the earth were tripped by the rapidly accelerating fuel pumps, throttling-up engines, and millions of pounds of gathering thrust, the first mission to leave the earth and visit another world began.

Paul Haney was steady as a rock, *"3, 2, we have commit. We have liftoff. Liftoff at 7:51 a.m. Eastern Standard Time."*

In the trench no one could look up because they were too busy. Behind the engineers in their seats Wernher von Braun was pacing, but pausing to look carefully over the shoulders of the men that were coaching his dream come true. Kurt Debus was a stoic mask as he looked over the booster figures pouring across Les Geisman's monitor.

"Booster is Go Flight," Geisman began and the reports were rolling down the trench.

"Retro is Go Flight,"

"Fido is Go Flight."

"Guidance?"

"We're Go Flight," Darrell didn't want to commit too fast. It was too early in the flight.

Paul Haney: *"We have cleared the tower."*

Outside the blockhouse, the nearby viewing stands were packed, and all along the beaches from Titusville to Cocoa Beach, early morning crowds of spectators watched as the fiery tail of Apollo-8 rose above the coastline and climbed into the Florida morning. The bright orange tail of the rocket flame could be seen clearing the scaffolding of the gantry works well before the wall of sound arrived and began pounding into the chests of the onlookers. When the sound wave came and surrounded the thousands with thunder, the rush of the emotion that followed was for many like nothing they had ever experienced. There were drowned out bursts of laughter at the sheer excitement, there were spontaneous tears produced by the absolute effect of glowing hearts, but within every American that stood and watched, there was a swelling of pride as the climbing Saturn-5 tipped slightly toward the ocean and rapidly continued to rise. No nation in the world had ever accomplished anything like this . . . and it was more than obvious to the multitude that was watching.

Inside the blockhouse the earth-quaking vibration was just passing the peak as the intercom speakers crackled into life with the voice of astronaut Mike Collins from Houston. "Apollo-8 you're looking good."

For the first trip to the moon Mike was the Cap-com for the most dangerous spaceflight to date. If not for a slipped disk and surgery on his back, Mike would have been aboard Apollo-8.

After a very long three-second pause, Frank Borman replied slowly with the single word, "Roger," and it was suddenly obvious that no one had ever ridden anything like the Saturn-5. The Redstone had been a launched pogo stick compared to this, the over-pressurized Atlas scary, and the Martin-Titan a G-force express train, but when a veteran astronaut and test pilot like Frank Borman took more than a second to answer, the conditions on the Apollo-8 flight deck had to be beyond anything expected.

"Approaching maximum high 'Q,'" Darrell said the words and everyone held on tight. Maximum 'Q' began at 78 seconds into the flight and was the most dangerous part of the ascent into orbit. It was when the dynamic pressures were at the most extreme because of the relatively still-thick atmosphere—at 44,000 feet—and it was a time when the entire stack of three stages was passing through the violent turbulence before breaking through the sound barrier.

Meanwhile, outside the blockhouse and in the viewing stands and all along the beaches, the crowds briefly observed a perfectly circular white cloud form around the upper center of the climbing rocket as the Apollo-8 astronauts slipped past the demon guarding the sound barrier and the terrible force of maximum dynamic pressure.

After maximum 'Q,' Darrell knew that von Braun was behind him and looking over his shoulder. The flight angles were unfurling perfectly. The actual was rising flawlessly beside the theoretical, and when the German passed to the next station in the trench, everyone knew it was time to breathe again at least for a few more seconds . . . at least until staging.

Before anyone could speak or even give thanks for passing through Maximum 'Q,' Mike Collins was on the job again and crackling over the speakers, "Apollo-8 . . . you are Go for staging."

After another 2-second pause and big Saturn's first stage relentlessly burning through another 30 tons of fuel and gaining an accelerated velocity of almost five times the speed of sound, Frank Borman answered evenly over the unsurpassed vibration, "Roger. We are Go for staging."

Chapter 92

"The first seconds of flight were a total surprise to everybody because the Saturn-5, which is a big tall rocket and we were like a bug on the end of a whip. It gets very massive at the bottom, with the center of gravity near the bottom, so if you rotate it, that little wiggle at the bottom translates into a big wiggle at the top.

For the first ten seconds — and it seemed like forty — we could not communicate with each other. Had there been a need to abort detected on my instruments I could not have relayed that to Borman."

— Bill Anders

When the real vibration and acceleration began, Bill Anders knew he was in the presence of heroes. No amount of training or simulator experience could ever prepare anyone for the feeling of a launch vehicle at liftoff . . . but it was something that the other two pilots had already been through.

Everything was right down the middle until the engines started and the jarring vibration began. The beginning of the explosion was the sound waves bouncing off the ironworks of the gantry tower, but when the whole stacked column suddenly gathered strength and tore its self from the sheer bolts and the massive release clamps, Anders suddenly knew why NASA had named this ground shaking monster after the Greek God of light.

As the unthinkable pressure and G-force continued to build above the continuous explosion from below, Astronaut Anders suddenly added another new thought to his first revelation about spaceflight: No one could ever be prepared for what was happening inside a command module during a booster launch unless it had happened before . . . but not even the most experienced veterans . . . had ever been aboard the most powerful rocket ever assembled.

When the battering ram of sound and shudder intensified beyond anything conceivable and followed the flash of light that took away the sunrise streaming through the only window not covered by the blast cover, the reactions of the two men that shared the cabin were almost identical; they were prepared as anyone could be, but it was obvious as Borman hit the elapsed flight time clock and the indicator began spinning, that no one could ever really be prepared for the seconds that somehow kept happening.

Jim Lovell was no longer an astronaut but a refined aeronautical instrument, when his words were scrambled by the earthquake that was rising, "We are clear of the tower . . . azimuth angle . . . good at . . . 72 degrees."

When the cabin rolled away and it was heads down as the actual acceleration began, there was a single thought that kept breaking through Anders' training. Watch these guys—they've done this before—even if it was Gemini— and not this monster. Watch for their reactions . . .

But the G-force pressure began building faster than any simulator could ever come up to speed, and Mike Collins was a distant blur through the pounding sound waves as he said ironically: "Loud and clear." And then Borman's voice was somehow slow and miraculously audible as he said, "Beginning roll and pitch program."

It was after that fleeting second, during the beginning of the roll and pitch program, that the destructive tremors took a new turn for the worse. It began as an unstoppable and unpredictable movement that rose up from the bottom of the stack to resonate in the entire framework of the spacecraft. As the G-force pressure continued to build at a crippling pace and the mounting vibration eradicated all sense of sanity and continued to escalate, it was suddenly obvious that none of the instruments were readable and the sound waves too intense too distinguish any words. Yet Borman and Lovell were impossibly going through the motions and the checklists just as in the simulator; Borman's hand, however, was firmly holding the D-ring abort lever.

Jim Lovell looked over with the warning, but his face and words were a blur in the thunder, "Coming up on max 'Q.'"

Through the continuous eruption of detonating fuel, Mike Collins was marking the passing of seconds with "Mark, Mode-1 Charlie," and Bill Anders was suddenly called back to focus on his training. 'Mode-1 Charlie' was just one of many abort codes. There were abort codes and procedures for every stage of liftoff—for every stage of the flight.

Before another thought could be crowded into the experience, Borman answered the call from Houston. His voice suddenly sounded strong and unwavering through the headset as he said, "Mode-C," but Borman's hand was the only steady object in the cabin as he was continuously holding the abort lever to launch the emergency escape tower—the abort lever that surely must be pulled to lift the capsule free of the exploding booster that *had to be* flying apart.

When Mike Collins' next words came cresting over the continuous roar, the meaning seemed unbelievable: "Apollo-8, Houston. You are Go for staging."

After another careful pause, and another two seconds passed into eternity, and another thirty tons of fuel exploded at 5,000 degrees through five of the largest rocket engines ever created, Borman replied smoothly, "Roger we are Go for Staging. Inboard engine is out at this time." With those words, Anders suddenly realized he could see through the vibration to the blurred launch-vehicle panel and the centermost engine light that had just winked out.

The next declaration that could be deciphered was Frank Borman saying, "Staging."

Seconds after the word, the exploding and ripping bedlam was over, and an abrupt but ominous silence followed as all three men were suddenly thrown forward against their straps as the G-force pressure was abruptly reversed. As the ominous seconds of silence slipped by and the pressing G-forces were relaxed, a new adventure in acceleration began with an unbelievably abrupt push and pressure when the five J-2 engines of the Saturn-5's second stage burst into life.

Once again, with a dynamic pressing G-force weight that could never be reproduced in a simulator, all three of the pilots were suddenly thrown backward and pressed down into the survival couches.

For a few brief seconds, a bath of orange firelight flashed and appeared in the hatch window before Frank Borman reached forward as the commander and triggered the escape tower release switch. Within a split-second an additional tremor announced the escape tower rocket's launch as the remaining windows of the spacecraft were abruptly uncovered. The escape rockets had fired with the actuator and pulled away the protective blast cover from the capsule.

When the blast cover and escape tower were away, an unbridled torrent of sunlight began streaming into the cabin as Bill Anders, for the first time, saw the sun and the mother of the earth without the filter of an atmosphere.

Away from the unshielded outburst of the nearest star, and the contrasting deep darkness of space, an unmistakable curve of blue oceans and white clouds began to appear in the extreme right corner of the fifth Apollo viewport. Across the distance between the dazzling sun and the rising curve of marbled oceans and clouds was the very black gulf of never-ending space.

Unexpectedly, the situation was obvious and a new lesson was learned. There could be no stars visible with the dominating sun crossing though the window ports and the solar glare as bright as the unshielded flash of a welder's arc.

Looking quickly across to Borman and Lovell bathed in the glaring sunlight, Anders suddenly realized that the second-stage acceleration and vibration was moderating down to almost like what the simulators actually forecast during launch vehicle training.

Amazingly, Jim Lovell was still going through the checklists and switch panels as rote when Borman's voice sounded over the headset. "Houston?" he transmitted from the left seat, "How do you read? Apollo-8."

During the overwhelming first stage ascent, Frank's responses to Collins at Houston had been guarded and measured, but the Cap-com answers were always immediate. "We hear you loud and clear Apollo-8."

"Okay," Frank Borman replied calmly, "the first stage was very smooth, and this one even smoother."

From his left side Bill Anders felt a jab and an elbow nudge and turned to see Jim Lovell's grin as Borman downplayed the violence of the Saturn-5 ride.

After the slightest pause, Collins reaffirmed, "Understand smooth and smoother. Looks good here . . ."

"Apollo-8, Houston," the confident voice from the astronaut in Texas continued, "your guidance and trajectory are Go. Over . . ."

"Thank you Michael."

"Yes," Collins answered, "you're looking real good Frank."

"Very good," Borman's tone sounded as if he were speaking to a head waiter and ordering dinner.

Outside the window viewports, the glare of the sun was being replaced with the reassuring curvature of the earth, and the edge of the atmosphere as the blue-and-white marbled aquamarine rose to touch the blackness of space. As the sun began to rotate out of view, the first of the stars began to shine and then grow brighter. The roll and pitch program was happening. It was right down the middle.

"Apollo-8 Houston," Collins reported, "your guidance and trajectory are Go—second stage still looking real good."

"Apollo 8's Go," Jim Lovell reported as he looked to Bill Anders.

Finding his voice and checking a figure on the console he could finally read, Anders replied calmly and was very proud of the tone of his voice, "Onboard chart confirmed. Check list Go."

"Roger," Collins answered smoothly. "Understand. You're looking good. "

Before there was even a second to celebrate, a new vibration started that was somehow different, but alarming in a different way. It was a new sensation beginning that was once again shaking the cabin and the flight deck, but this was a bouncing upheaval that seemed to oscillate and gather strength in rising intervals. It was a different feeling from the violence of the first stage, but the quick glances that Borman and Lovell gave to each other offered another alarm signal that was never practiced in a simulator.

Mike Collins was back in the headset with more reassurance, "Apollo-8, Houston. Your guidance and trajectory are Go. Over."

"Roger," Frank came back during a peak in the mounting vibration, "we're picking up a slight pogo at this point."

"Understand," was the smooth Cap-com reply, "slight pogo. Thank you."

Without waiting for another response about the pogo, Mike Collins was once again the determined and steadfast coach. "Apollo-8," he transmitted, "you look good for staging."

"Roger, staging," Borman replied. "Pogo is dampening out, we're right down the middle."

Meanwhile down on the ground, in the trench, and all across the viewing stands and broadcasting over the airwaves, Paul Haney continued with his official running commentary.

"We're 7 minutes into the flight, and we're nearing the second stage — nearing the point where we'll drop the second stage and light the third stage. That event is to come at 8 minutes and 40-odd seconds into the flight. We have now achieved 70 percent of the velocity required to obtain orbit. Our present velocity is 18,600 feet per second and we're 100 miles above the earth. 100 even."

Chapter 93

*"When Deke Slayton spoke to Jim Lovell's wife Marilyn and told
her the Apollo-8 mission had about a fifty-fifty chance of making
it to the moon and bringing her husband back, she replied: 'Oh
that's good! Jim said he probably had only about a thirty-percent
chance of making it home.'"*

In the Trench - 11 Minutes After Liftoff

With the white smoke of the rocket contrail beginning to
drift above the rising sun and the early morning December
breeze, the voice of Paul Haney continued over the outside
viewing stand loudspeakers and most of the radio and television airwaves around the world.

*"At ten minutes and fifty seconds into the flight, we've heard
from Bill Anders for the first time and he received a 'looking good'
comment from Mike Collins."*

"Come on Hp-2, come on . . ." Darrell said the words
firmly but quietly. He was a rock as he stared at the guidance information streaming in before him. Von Braun and
Kurt Debus were standing behind the Cowboy along with
Les Myers and two of the engineers from IBM. The bear that
was Les Geisman was anxious in the next seat, as was everyone else in Houston and at the Cape that were also willing
the Saturn's third-stage booster to complete the velocity and
height requirements for orbit. Apollo-8 was at the tip of a
hundred tons of hardware that was hopefully headed for
the moon and a good story in the history books.

Everyone watching the monitors knew that if the "Height
of the perigee — Hp" was not higher than the curvature of the
earth and the engine failed before the Hp-2 figures unfurled

on the monitor, Apollo-8 and the first three Saturn-5 astronauts would free fall back to earth. There was, of course, escape, abort, and reentry plans for every stage of the flight, but the first mission to the moon would be scrubbed, and there would be finger-pointing congressmen that would go crazy with: "I told you so!"

"Hp is coming up," Jim Lovell reported smoothly over the speakers. "The S-4B is burning perfectly. Hp is now plus . . ."

"Thank God for that," Les Geisman responded for everyone watching.

"Ja, Ja, wonderful!" Kurt Debus exclaimed. "Excellent!"

The third stage fired perfectly during the last episode of staging and with the latest report of Hp-2, Apollo-8 was now going into orbit. The contingency plan before this level of flight was to drop a failed S-4B booster and fire the engine on the Apollo service module to gain a low earth orbit. It was an official plan, but not a favorite of anyone in the trench. If the S-4B failed there would be no trip to the moon.

"Borman and Lovell are amazing," the flight surgeon looked up from his screen and leaned back in his chair. "Their heart rates and breathing are almost back to normal and hardly accelerated, but our new boy Bill . . ." the surgeon grinned. "He's having quite the ride."

After H-p2, there was a moment to actually breathe and time for a quick look around, but everyone that had heard the doctor's remarks had a lopsided smile.

"*This is Apollo control here,*" Paul Haney was once again the polished voice of reason sounding over the speakers. "*We are 21 minutes, 44 seconds into the flight and we are in a near perfect and circular orbit of 103 miles and coming up just over the Canary Islands. The communication with the spacecraft has been nothing short of outstanding. I don't recall a time of when the communication from a simulator was this sharp and this clear as it is today from this spacecraft. Here's how the conversation is going as we proceed across the Canary Islands.*"

"Apollo-8 Houston," Mike Collins was transmitting, and tension in the control room was beginning to settle down.

Kurt Debus and von Braun, however, were on the move again and marching along the monitors.

"Go ahead Houston, Apollo-8," Frank sounded clear over the Canary Islands with an obvious growing confidence.

"Roger," Collins was quick to come back, "you have less than one minute to LOS. Everything is looking good on-board the spacecraft and the S-4B. We will pick you up again over Tananarive in 37 minutes."

"Roger," Borman acknowledged officially, "thank you Houston, Apollo-8."

"How about let's take off our helmets and gloves?" Jim Lovell suggested, and his voice was very clear over the speakers and the still-active AOS link. "I mean, let's get comfortable," he said. "This is going to be a long trip."

Without delay Bill Anders answered, "Okay," and Commander Frank Borman agreed, "Good idea," just before the crackle of LOS.

Around the control room there were more lopsided smiles, nodding Germans, and the pretty Pan-Am flight attendants that knew everyone present was working on the most important project in the world — the most important project ever.

Darrell took a moment and thought about what was happening. There were three Americans on their way to the moon.

Chapter 94

"By the way, if any crackpot says we really never went to the moon, we were measuring the distance to the spacecraft the whole time, so we know they went to the moon."

From the archives of Hamish Lindsay and the tracking station crew at Honeysuckle Creek, Australia.

Mid-Morning, December 21, 1968 – Titusville, Florida

After the Apollo-8 launch and the biggest mission ever was up and away, Audrey knew she needed to get ready for Christmas. She had watched the beginning on television and then gone outside with the kids and all the neighbors and watched as the unbelievable Saturn thundered up and rattled into the sky.

The sight and the awesome sound was as always inspiring and wonderful, but terrible to watch. Inspiring because it was the greatest thing that had ever happened, and terrible because of what could go wrong at any half second. If just one tiny part failed, Audrey knew it would dash the hopes and dreams of all involved, and of everyone else in America who needed desperately for the mission to the moon to work.

As Audrey watched the rocket contrail begin to drift and fall apart and the neighbors beginning to file back into their homes, she thought once again about the terrible consequences if something happened and the three Americans were lost in space. Loaded with her latest worry, Audrey followed the children back into the house and decided that this year she was going to go all out for Christmas. If something happened to the moon flight, she knew that Darrell and the kids would be devastated, and if the house was decorated everywhere for the holidays, it might help to take away some of the pain. Christmas was her favorite holiday and in the past, it had helped her cope when Darrell was away. Now, she hoped it would help if the unthinkable happened.

With Mike and Deb glued to the T.V. and little Kath fidgeting as she sat watching Walter Cronkite broadcasting from the Cape, Audrey became more convinced than ever she had to get busy with Christmas. She needed it and so did the rest of America.

The beginning of the year had started out rough, and with every passing week there was only more bad news. The way things were going it couldn't get much worse — unless of course — something happened to Apollo-8.

Vietnam was a mess and the nightly reports with gunshots, explosions, and Americans killed in battles halfway across the world was weighing everyone down. And with the Russians backing the Communists, no one really knew what was going to happen. It was the old Atomic Age creeping back again with new and terrible details waiting in every newspaper.

Mike had come home from school one day and told the story about a boy in class that had asked the teacher a question. He had asked: "Why can't we drop an atomic bomb on North Vietnam like we did on Japan? It ended that war, so why can't we end this one the same way?" The teacher had answered: "Because Russia won't let us."

Audrey had been troubled by the story and she knew that all the children in the class had too.

As long as Apollo-8 worked as planned, it was taking everyone's mind off of Vietnam, the protesting mess in Washington, and the unthinkable atomic bombs on the tips of Communist rockets. But then she considered that the killing of Martin Luther King and Robert Kennedy had really thrown the year into a tailspin. The Civil Rights leader was shot and killed in April, and before the month was out, the race riots that followed had burned down buildings all across the country and thousands had been arrested in Washington.

Just as the civil rights killing in Memphis was beginning to settle down, Robert Kennedy had been shot to death in California. He had promised to end the war in Vietnam and to stop the killings of hundreds of Americans every week overseas. He had also promised to pave the way for new breakthroughs in civil rights.

It was hard to believe that so much could go so wrong in just one year, and with all of the soldiers dying every day in Vietnam, everything had seemed so hopeless until that wonderful night with Darrell, the kids, the rocket, and the moon rise, and now it was almost Christmas and three men were riding one of those rockets — and they were headed for the moon.

From the living room Audrey heard the kids arguing. Deb and Mike wanted to watch Walter Cronkite and the flight to the moon, but Kath wanted to watch a western. Kath loved her horses.

When the older kids had convinced Kath that the moon coverage was better because that was what daddy was working on, Audrey went into the garage and then up and into the attic. Thoughtfully, she began gathering the decorations for Christmas. This year she hoped Christmas would be perfect, but she was very afraid it wasn't going to be.

Chapter 95

"Oh give us the man who sings at his work."

— Thomas Carlyle

December 21, 1968 - Aboard Apollo-8, at 24,500 miles per hour

"Fly me to the moon . . ." Jim Lovell was singing, "and let me dance among the stars . . . Show me what it's like . . . on Jupiter and Mars. In other words . . . hold my hand . . ."

"Those are not the right words," the young astronomer with the doctorate degree and the black horn-rim glasses said as he stood up. "Those are not the right words! It should be: *play* among the stars, and show me what *spring* is like on Jupiter and Mars."

"Oh brother," Les Geisman shook his head after a glance to the gangly young man with the rolled-up shirtsleeves and the pocket protector stuffed with mechanical pencils and fountain pens.

Darrell snorted as he laughed. Kurt Debus and von Braun were clinically examining the young astronomer as if he had three heads, and half the others in the control room either laughed or rolled their eyes at the comment made by the eager perfectionist.

"I'm glad that's all we've got to worry about for the moment," Les Myers voiced the comment and stopped to look over Darrell's shoulder.

The critical trans-lunar injection burn had lasted just under 7 minutes, and after a two-hour parking orbit, the third stage of the Saturn-5 reignited and rapidly boosted the Apollo-Saturn combination from a 17,400 mile an hour velocity to an unprecedented speed of just over 24,000 miles an hour. No human being had ever gone as fast, and Jim Lovell was singing happily as if he were in the shower. He had a great voice that was loud and clear over the intercom.

"It's official," Darrell said from his seat in the trench. "They just broke the altitude record from Gemini-11."

"Pete Conrad and Richard Gordon were the first to go this far out, and the first to see the earth as a complete sphere," Les Myers recalled. "There was a lot of pioneering done with Gemini-11."

"We're going to be breaking a lot of records on this flight," Les Geisman offered. "These guys are really on the way to the moon. Even as much as we've practiced and waited, it still seems unbelievable."

After the critical guillotine-style separation and the explosive bolts fired separating the command and service modules from the Saturn-third stage—or what the technicians called the S-4B—the Apollo-8 capsule and service module spacecraft were now surging forward and onward toward the moon. The separated and much larger S-4B booster, however, was also traveling at lunar insertion speed and would be a collision hazard until an additional separation maneuver could be completed. There were also four jettisoned and tumbling panels headed to the moon that would have covered the spider-like lunar excursion vehicle had the landing craft been ready.

Over the intercom, the signal from Apollo-8 was strong. "We see the Earth now," Frank Borman reported, "almost as a disk."

"Good show," Mike Collins answered jovially from Houston. "Get a picture of that."

"We are, and by the way," Frank sounded as if he were smiling, "tell Conrad he just lost his record for Gemini-11."

Before the Cap-com from Texas could answer, Jim Lovell's voice broke over the speakers. "We have a beautiful view of Florida now. We can see the Cape. At the same time we can see Africa. West Africa is beautiful. I can also see Gibraltar at the same time I'm looking at Florida."

Collins replied, "Roger. Sounds good. Take lots of pictures. Are the windows clear?"

"Roger," Lovell responded and there was wonder in his voice. "I can see the entire earth now out of the center window. I can see Florida, Cuba, Central America—in fact, I can see all the way down to Argentina and through Chile."

Collins replied softly and everyone could hear the pride, "They picked a good day for it."

Suddenly Borman was back on the intercom and all business, "Stand by one." He transmitted from space. "We are going through the separation maneuver checklist. We've lost sight of the S-4B here. The separation maneuver may be delayed slightly, or else we will have to do it without having her in sight."

"Roger, I understand Frank." Mike Collins might have been in Houston, and on Planet Earth, but his heart was with the three men who had just crossed a boundary and were further away from humanity than anyone had ever been before.

"When does the S-4B do that blow-down procedure?" Borman was worried about the Saturn third stage, just as he had been concerned about the whereabouts of the ghostly and persistent Titan-2 second stage on Gemini-7. No one wanted to fire off a thruster and hit something that was not visible from a viewport window.

"Stand by one . . ." Mike Collins and the crew from Houston were checking the figures.

Les Geisman stretched in his chair and shook out a Chesterfield. Darrell shook his head and smiled as he thought about the three men that were able to see the entire earth from space.

Across the control room, Wernher von Braun and Kurt Debus were pacing and the girls from Pan-Am were making another run.

"The Russians never got off the ground, did they?" Geisman lit his cigarette and blew out a smoke ring.

"As far as I've heard they never did," Darrell looked thoughtful. "This would have been it too," he said, "the first men to the moon. The Russians had their Zond-7 spacecraft ready to do a manned lunar orbit. The whole program was lined up just like ours and planned for earlier this month, but they canceled at the last minute. I guess their N-1 super booster was just too unreliable."

"Let's hope everything we've got is still reliable," the big Texan offered. "But without those two," Geisman said as he tilted his head toward von Braun and Kurt Debus who were now heads down and in deep conversation, "we probably wouldn't be sitting here today."

"There's no doubt about it," Darrell agreed. "You can't get pilots to the moon without lifting the payload it takes to get them there, and the Boss's numbers were always right down the middle."

Chapter 96

*"We came all this way to explore the moon, and the most impor-
tant thing is . . . We discovered the earth."*

— Bill Anders

Christmas Eve, 1968

"This is it," Les Geisman whispered. "We'll know in a
few minutes whether we're making a historical milestone
of men to the moon or making widows out of some women
here on the earth."

Apollo-8 had disappeared right on schedule behind the
dark side of the moon and was now in LOS. Borman, Lovell,
and Anders were now traveling backward — or engine bell
forward — and in a slightly heads-down or capsule-down
position as the spacecraft crossed 86 miles above the crumbling craters, the gray oceans of dust, and the uneven and
broken horizon on the lunar surface.

The plan on the big board was to fire the service module propulsion engine for a 4-minute burn to slow down
enough to enter lunar orbit. If the engine failed to ignite, or
the full 4-minute burn wasn't completed, Apollo-8 would
still come out from behind the moon but perhaps be lost in
an everlasting orbit somewhere between the earth and the
eternal lunar surface. If the engine burn lasted too long, or
if the astronauts were unable to shutdown the retro maneuver, the three Americans would crash onto a desolate and
unforgiving alien landscape.

Darrell had never felt more stress and pressure, and he knew he was not alone. The survival of the three astronauts depended on the skill of the pilots, the hardware on the dark side of the moon, and the exact calculations from the technicians in the trench. If the propulsion engine did not fire, Apollo-8, with Borman, Lovell, and Anders, would reappear as a radio signal in 3 minutes. If they did not, they were either crashed into the unseen and Russian-named Moscow Sea, or they would appear to the earth-bound antennas 22 minutes later and would have successfully arrived in a stable lunar orbit. That, at least, was the plan that had been scrutinized over and over again.

Once again, the official public relations voice of Paul Haney sounded over the speakers and added to the clicking teletypes and the multiple conversations from the trench and the crowded control room.

"This is Apollo Mission control. At 69 hours and 8 minutes from liftoff Apollo-8 is 30 seconds away from the time of planned lunar orbit insertion burn. The crew should now be looking at the countdown to ignition on the face of their display and keyboard. In a slightly heads-down attitude they should also be looking down over the rugged features of the back side of the moon moving below them at a high rate of speed. Standing by, this is Apollo Control."

"It's either widows and orphans or heroes," Les Geisman said quietly and then looked at his watch. "We'll know any minute. The worst thing would be if they just disappeared and we never heard from them again. It could kill the program."

"Anything could happen at his point," Darrell conceded, "but the math and physics are strong. One way or the other, they'll come out from behind the dark side."

"The real question is," Geisman was looking down the trench, "whether or not that service module engine will fire and bring them back home. If it doesn't work now it probably won't work at all. There's a lot riding on one engine that's 240,000 miles away."

"This is Apollo Mission Control. Mark, one minute from predicted time of acquisition of signal."

"Apollo-8 Houston," today the voice calling from Houston was Jerry Carr as Cap-com.

Once again everyone in the trench was beyond focused. The sheer will in the control room had to be a tangible energy. Debus and von Braun had stopped their pacing and were staring at the overhead intercom speakers. There was a lonely and very deep static as the new squelch levels searched for a voice from the moon.

"This is Apollo Mission Control, we've acquired signal and telemetry but no voice contact yet. We are looking at engine data, and it looks good. Tank pressures look good. We have not yet talked with the crew but are standing by . . ."

"Apollo-8 Houston," the voice from Texas continued to call.

With an abrupt crackle, the loud and clear voice of Jim Lovell broke over the speakers, "Go ahead Houston. This is Apollo-8. Burn complete, our orbit is 169.5 by 60.5."

"Apollo-8, this is Houston," Jerry Carr related as everyone in two control rooms began to breathe, beam, laugh, and clap their hands. "It's really good to hear your voice Jim."

"That's amazing," Les Geisman shook his head, "that AOS came in at the exact second that it was predicted by the numbers."

Darrell grinned, "Right down the middle."

Chapter 97

CBS anchor Walter Cronkite about the first Apollo moon flight:
"Everything else that has happened in our time is going to be an
asterisk."

Christmas Eve - 60 Miles Above the Moon in Lunar Orbit

"This is Apollo Control Houston," Paul Haney was once again sounding over the speakers in the Cape Kennedy control room. *"We are 75 hours and 47 minutes into the flight and are due to acquire AOS at just any second by a whole host of earth-bound stations. This is the fourth revolution around the moon by a manned spacecraft."*

"Apollo-8 this is Houston," Mike Collins was back at the Cap-com station and transmitting from Texas. "Apollo-8 do you copy? . . . Over?"

"Okay Houston," Jim Lovell was acknowledging the contact with a fresh burst of static as Frank Borman gripped the flight control lever and rolled the spacecraft to line up the high-gain antenna. Without a proper alignment, the VHF and the high-gain transmissions were sketchy. When the alignment was good, the transmissions were perfectly clear. For the beginning of this latest AOS, the reception and alignment had been terrible.

After the extended quad-dish antenna finally locked in with a near perfect signal strength and Apollo-8 stabilized, Borman held the flight control lever steady as the command module was pitched downward. For the pilots in Apollo-8, this new and latest position was providing the best and most scenic view of any of the lunar orbits.

With Bill Anders clicking away with the 70-millimeter lens on the coveted Hassleblad camera, Lovell began a running description of the lunar surface as the mountains of the moon passed below.

"The moon is essentially gray, no color, looks like plaster of Paris or sort of a grayish beach sand. It really looks like dirty beach sand. We can see quite a bit of detail."

Across the control room two of the technicians and two of the Pan-Am girls were decorating a small Christmas tree. The teletypes clattered in the background and the cigarette and cigar smoke was as thick as the anticipation that filled the room.

"The Sea of Fertility doesn't stand out here as well as it does on Earth," Lovell continued over the intercom and the reception was improving and near perfect. "There's not as much contrast between that and the surrounding craters. The craters are all rounded off. There are quite a few of them. Some of them are newer. Many of them look like—especially the round ones—look like they have been hit by meteorites or projectiles of some sort."

Darrell was munching a club sandwich in between sips of hot coffee. He wanted to go outside for a breath of fresh air, but it wasn't everyday that voices from the moon were describing a play-by-play flyover of another world. The cigarette smoke was bad enough but the cigar smoke was worse.

"Langrenus is quite a huge crater," Jim Lovell continued, and everyone could imagine him looking down and out of a window as the lunar landmarks passed below. "The crater has a central cone to it. The walls of the crater are terraced, about six or seven terraces all the way down. That one dark hole . . ."

"Oh my God!" Bill Anders voice suddenly broke in over the speakers. "Look at that picture over there! Here's the earth coming up. Wow, is that beautiful!"

"Hey, don't take that," Frank Borman scolded. Then he laughed. "That's not scheduled," Borman said as if he were chiding Anders for taking unauthorized photos.

"Hey give me that color film—quick!" Anders called out from 240,000 miles as he was fighting to get his film ready. He was the first human ever to see an earth rise, and as he rushed to prepare his camera, Borman and Lovell sat transfixed as they too were spellbound as the earth began to climb away from the mountains on the moon, the gray shadows, and the broken cinnamon rim on the lunar horizon.

Across the gulf of space, the blue oceans of life were a splash of color as streaming white clouds crossed over the dark green jungles and the tan and sandy continents. Rising six times bigger on the moon than could be seen at home, Mother Earth with daylight shining easily dominated the gray oceans of dust and the ancient craters of a timeless world.

For a solid moment everyone in the control room froze— the Pan-Am girls beside the Christmas tree, the technicians sitting in their chairs, and even Kurt Debus and von Braun— as they stopped all motion and concentrated on the unexpected words that were coming from men in a spacecraft that was orbiting the moon.

"Man that's great," Jim Lovell said and his voice was filled with awe. "Oh, that's a beautiful shot. You sure we got it now?"

"Yes," Anders' voice cracked. "We'll get—we'll—it'll come up again, I think."

"Just take another one Bill," Lovell coaxed gently, and everyone from Houston to the Cape and all the way to Honeysuckle Creek in Australia could hear the emotion of what the first earth rise was like as it slowly rose from the dusty oceans and the desolate lunar mountains.

To the first three humans that ever traveled sixty miles above the moon, the emerging earth was a blue-and-white marbled sphere shaded slightly with nightfall as it continued to climb away from an unknown and unforgiving landscape.

Darrell shook his head in wonder, "I don't think anyone ever expected that, to watch the earth rise over the lunar horizon. It must be huge compared to what the moon looks like to us here on earth."

"Well," Les Geisman gave a crooked grin, "I never expected to be here, helping our side circle the moon when the Russians launched Sputnik twelve years ago."

Darrell leaned back in his chair. "I can't believe it's only been twelve years," the Cowboy shook his head. "So much . . . we've learned so much . . . and so much has happened in only twelve years."

Chapter 98

"The vast loneliness is awe-inspiring and makes you realize just what you have back there on the earth."

— Jim Lovell

Christmas Eve - Above the Sea of Tranquility

"This is Apollo-8 coming to you live from the moon," Bill Anders was suddenly on television in the command module cockpit and it was obvious that he was weightless as he gripped a handrail. Jim Lovell was filming and he had chosen a perfect angle. Over Anders' shoulder was a triangular viewport window with the craters of the moon passing below.

Les Meyers spoke softly over the transmission but everyone in the trench could hear. "It's official," he said, "there are more people watching this broadcast than at any other time in television history. Sixty-four countries are airing this on radio and T.V."

"We are now approaching the lunar sunrise," Bill Anders continued, "and for all the people back on earth, the crew of Apollo-8 has a message that we would like to send to you."

There was a brief pause and then the words from the King James Version of the Bible began. Anders voice was perfectly clear as he began to read from the Old Testament.

"In the beginning God created the heaven and the earth. And the earth was without form and void, and darkness was upon the face of the deep. And the spirit of God moved upon the face of the waters. And God said, Let there be light: and there was light. And

God saw the light, that it was good: and God divided the light from the darkness."

Above the moon and aboard the spacecraft the lunar darkness had just fallen away as Apollo-8 traveled into a new orbital sunrise.

In the trench and throughout the control room everyone was suddenly drawn into complete silence as they sat in reverence and listened. After a crackle of static and a change of voices, Jim Lovell continued reading from the first chapter of Genesis.

"And God called the light Day and the darkness he called Night. And the evening and the morning were the first day. And God said, let there be a firmament in the midst of the waters, and let it divide the waters from the waters. And God made the firmament, and divided the waters, which were under the firmament from the waters that were above the firmament: and it was so. And God called the firmament Heaven. And the evening and the morning were the second day."

Frank Borman continued the verses without interruption. His voice as a commander had never sounded stronger.

"And God said, let the waters under the heaven be gathered together unto one place, and let the dry land appear: and it was so. And God called the dry land Earth; and the gathering together of the waters called he seas: and God saw that it was good.

"And from the crew of Apollo-8," Borman finalized, "we close with good night, good luck, a Merry Christmas — and God bless all of you . . . all of you on the good earth."

When transmission ended and once again Apollo-8 began the hour-long journey around the dark side of the moon and the impervious LOS, Les Myers shook his head.

"I've read the flight plan over and over," he said, "and I never expected that. For the last television broadcast all that was official in the NASA flight instructions was to "do something appropriate.'"

The duration of an Apollo-8 lunar orbit at 60 miles above the moon was just over two hours. One hour was spent in LOS behind the lunar surface and one hour crossing front when the high-gain quad-dish antenna could lock on for AOS.

The three astronauts of Apollo-8 took photographs until they were exhausted and tried to keep up the flight plan and the rigid schedule that was assigned to them, but during the last two orbits, Commander Frank Borman ordered Lovell and Anders to sleep before the critical service module burn that would bring them back home.

When the service module engine did indeed fire as scheduled, and Apollo-8 achieved the speed required to leave lunar orbit and once again started streaking back toward the earth, Jim Lovell's first comment as the spacecraft emerged from behind the moon was a perfect Christmas gift for 1968.

"Roger," Lovell transmitted as Apollo-8 shot around the moon and was flawlessly headed back toward earth, "please tell everyone at home that there *is* a Santa Claus."

After another 240,000-mile voyage toward the blue-and-white marbled sphere that was the birthplace of humanity, Apollo-8 was bathed in a pale blue light — a light as bright as daylight through the window ports — as the three astronauts reentered the atmosphere and splashed down perfectly into the warm waves of the Pacific Ocean. It was the first time humans had visited another world and returned safely to the earth.

Chapter 99

"I wish you'd tell Dr. Von Braun, Lee James, Kurt Debus, and Rocco Petrone thanks a lot for all the people who worked on the great ride."

Tom Stafford on the flight of Apollo-10

"Charlie, my hat's off to the guys in the trench. I love them."

From Gene Cernan in Apollo-10 to Charlie Duke as Cap-com

"I'm telling you guys there's something buried under the moon," the youthful astronomer with the black horn-rim glasses was standing next to a flip chart and demanding attention like never before. "There's something big under the lunar surface," he insisted, "something enormous."

Darrell and Les had just arrived for their early morning shift in the trench and were discussing the latest reports for Apollo-10 when the expert on celestial mechanics began voicing his outlandish theory. The astronomer was doing his best to attract attention and hold court, and as usual with the crew in the control room, the concept of strange and unheard-of phenomenon was as always earning a few moments of careful consideration.

As the engineers and technicians filed in toward the monitors and the morning shift change, almost everything was right down the middle for the latest lunar mission that was carrying three astronauts to the moon. There were, however, more than a few issues as the Apollo spacecraft with the docked-and-attached lunar landing vehicle were crossing through the black sea of space and were headed spider-legs-first for the moon.

The gangly astronomer was determined as he stood his ground in front of his flip chart and the drawing he had carefully produced with multi-colored markers. "There is definitely something buried under the lunar surface," he nodded solemnly as he gave his warning, "something big enough to have an effect on the orbits of *any* spacecraft."

"The math doesn't lie," the astronomer was tapping a list of figures on his carefully drawn chart. "It has to be something massive to change the stable orbit of a spacecraft passing overhead."

On the astronomer's sketch, there was a circle with drawn craters representing the near side of moon and then several larger but oblong elliptical tracks that represented the now historical orbits of Apollo-8. There was indeed a marked difference in the drawing when the orbital track crossed over the flat areas that were the penciled-in lunar seas.

"It could be as simple as a giant iron meteor that crashed long ago and melted into magnetic pools under the surface, or it could be something else—something that we cannot even begin to understand. Something like an ancient spacecraft from far and away that was lost and buried under the moon."

After the latest celestial theory was explained, several of the technicians moaned and then began drifting back to their work stations or out and into the early Florida morning. A few of the others remained to examine the orbital mechanics that were the factual math and the unexplained deviations in orbit. The orbital anomalies were really not new news, but there had been vague rumors of ruins or decayed structures on the dark side that appeared ancient, and odd theories on why the dark side had mountains but the earth-side had only sandy oceans and vast craters.

"Imagine being halfway to the moon," Darrell said as he turned the big bear of a man that was Les Geisman away from the speculating astronomer and the ancient spaceship theories, "and not being able to sleep for days. Our guys are really excited to be doing what they're doing," the Cowboy continued, "and going where they're going, but they've got to be exhausted. Every time they're scheduled for a rest period, the reactive control thrusters are firing on the computer's command and the bumps and the noisy cycling is keeping them awake. If the two docked spacecraft travel even slightly off dead-band course, the RCS thrusters fire and wake everyone up."

"What about the drinking water?" Les Geisman asked pointedly as the two men gained their work stations in the front row of monitors. "Tell me about that."

"That drinking supply is way over-filled with chlorine," the flight surgeon interrupted from his seated position in the trench, "and every time they drink they get a mouth full of water that tastes like bleach or a swimming pool gone bad."

The surgeon shook his head and focused on the astronauts' bio-rhythms streaming on his monitors. One of the pilots was asleep and two were not.

Once again at the Cape it was business as usual: the three stages of the big Saturn-5 had performed flawlessly and lifted the 125-ton payload into earth orbit and then into trans-lunar-injection. The latest Apollo flight was once again headed for a manned mission the moon after Apollo-9 had successfully tested the Grumman-made lunar landing vehicle in earth orbit.

"On the next sleep period we should have the chlorine level right," the flight surgeon continued after he lit a fresh cigar and was puffing furiously. "We didn't know they'd taste that much chlorine in the water."

"Okay," Les Geisman scowled through the cigar smoke and drawled with his Texas accent, "they can't go to sleep and they have to drink water that tastes like its coming out of a poisoned well. I wonder what the boys and girls out there in T.V.-land would think about our futurist spaceflight now?"

"Don't forget the insulation explosion," Darrell voiced a major concern. "Those Mylar particles are everywhere and some of them so small the boys are having trouble keeping their noses clear. That's got to be really uncomfortable in a weightless and shirtsleeve environment. I'll bet that fiberglass is making them itch and scratch like crazy."

During the docking with the first lunar module headed for the moon, an insulation layer had exploded under pressure and tiny particles of fiberglass were everywhere between the landing vehicle "Snoopy" and the command module "Charlie Brown." In an instant, exploded fiberglass particulates were everywhere inside both spacecraft.

Les Geisman leaned back in his the chair and shook out a Chesterfield. He then took a sip of coffee and glanced at his telemetry information steaming on his screen and then over to the quirky astronomer where the gangly young man was still losing his observers. Les then looked over to the flight surgeon puffing on his cigar.

"Darrell," Geisman offered quietly as he blew a smoke ring and looked around the bustle of the control room. "I love this job."

The Cowboy grinned, "Me too."

Chapter 100

"At liftoff, I cried for the first time in twenty years, and prayed for the first time in forty."

Arthur C. Clarke, author of:

2001, a Space Odyssey after the Apollo 11, Saturn-5 launch

July 29, 1969 - On the Lunar Landing Module Eagle: 49,000 Feet Above the Sea of Tranquility

"They're not on VOX," Les Myers said with disbelief. "They're on 'Push to Talk.'" The control room superintendent was standing behind and in between Darrell and Les Geisman. His hands were on his hips and the tension in the trench had never been this high. It was as high as the men that were hovering above the lunar surface and were about to attempt a landing on the moon.

VOX was Voice Activated Transmission and push to talk was just that: Push to Talk. Neil Armstrong and Buzz Aldrin were clicking switches when they wanted to communicate and not clicking the transmit buttons when they wanted their conversation kept private.

The two men had recently separated from Mike Collins in the Columbia command module of Apollo-11 and were now standing side by side and shoulder to shoulder in the lunar landing module Eagle as they continued the orbital descent over the ancient craters and the boulder-strewn landscape below. The closer they came to the untouched debris field of otherworldly boulders and gray and cinnamon sand, the more uncertain they were of finding a place to land. There was only so much fuel in the landing and descent stage and they had to land the Grumman vehicle upright—if they wanted to ever leave again.

"Columbia should be coming up on AOS," Darrell checked his figures. "Mike will have line of sight and AOS three minutes before the LEM comes around the dark side. He will be able to relay any—"

Before Darrell could finish, Charlie Duke the Cap-com from Houston came over the speakers. "Columbia, Houston! We're standing by."

"Mike should be coming into range any second," Les Geisman whispered, "any second and we'll know if they're really going down."

With a fresh crackle of static the voice of Mike Collins appeared over the airwaves from a far corner of the moon. "Houston, Columbia," Mike reported through the gulf of space. "Reading you loud and clear, how do you copy me?"

"Roger, five by five, you are coming in perfect Mike," Charlie Duke sounded impressed, relieved, and then very curious. Everything depended on what happened in the next few minutes and everyone in the program was beyond anxious for the next words coming from the moon.

"How did the Descent Orbit Burn go?" Charlie asked, and everyone was frozen in their places and looking at the speakers.

"Listen babe," Mike reported with a smile in his voice, "everything is going just swimmingly—absolutely beautiful!"

"Great!" Charlie Duke responded. "We're standing by for Eagle."

"Okay, he's coming along."

It was Charlie Duke's turn to smile and everyone could hear the smile in his voice. "We copy . . . and Columbia . . . we expect to lose your high-gain transmission sometime during the powered descent."

Mike Collins was obviously in great spirits. "Roger," he said. "You don't much care, do you?"

"No sir." Duke's answer was swift and humorously honest. The limit of angle for Columbia's high-gain antenna to line up with the earth tracking stations was reaching a point that would be unattainable without an altitude change. Because it was vital to maintain secondary contact with the descending lunar module and Aldrin and Armstrong, Mike was compelled to stay in contact with the weaker but still effectual Omni-directional antennas.

Before any further transmissions were possible from the single astronaut orbiting 60 miles above the moon, the voice of Buzz Aldrin in the lunar landing module crackled through the speakers.

"Houston, Eagle. How do you read?"

"We copy you five-by-five Eagle." Duke and Aldrin sounded as if they were speaking over an inter-office intercom. The communications link was perfect. "We're standing by for your burn report."

"What are the figures Darrell?" Gesiman was leaning over and looking at the Cowboy's monitor. The telemetry figures were already streaming in with Aldrin and Armstrong suddenly appearing with the descending Eagle that was now in AOS.

"Looks right down the middle on apogee and perigee," Darrell answered as he double-checked the figures. "High orbit at 60 miles and low at 9, but they've burned a lot of fuel and 9 miles high is still 50,000 feet. They've already gone through a lot of fuel and they still have a long way to go. They've just slipped below perigee. Right now they're at 49,000 feet."

"Eagle, you are Go for powered descent," Charlie Duke said the words, and down the trench Kurt Debus sucked in this breath. It sounded to Darrell just like the day in White Sands when the blast wall of sand was suggested and the Redstone lost her guidance and plowed into the desert.

"Engine arm descent," Aldrin reported. He was now back on VOX. "We are 40 seconds until descent engine start."

Over Buzz Aldrin's voice-activated microphone, everyone in the trench could hear Neil Armstrong calmly ask as if he were a Florida tourist on vacation, "Is the 16-millimeter camera running?"

"Yeah," Buzz sounded over the VOX, "camera's running."

"Okay," Armstrong said and was heard over Aldrin's microphone.

"Override at 5 seconds. Descent is armed," Aldrin voiced the warning and it sounded over the VOX. "Altitude light is on."

They don't have a radar lock on the surface," Les Geisman offered worriedly. "That's why the altitude light is on. If they don't have solid radar lock, they'll not be getting the info for altitude and rate of descent."

Armstrong's voice was garbled and distant, but it was clear when he said: "Proceed with descent."

Aldrin sounded strong over the VOX as he began a quick an improvised countdown, "Proceed . . . One . . . Zero . . ."

"Ignition," Armstrong's voice was like a quiet but determined echo.

Aldrin answered, "Descent engine ignition. 10-percent thrust. Throttle up looks good."

Over the VOX, everyone in the trench could hear the subtle note of the descent engine and the popping of the pitch, yaw, and roll thrusters as the four groups of quad RCS cones popped out tiny bursts of fuel to control the flight angles.

"Okay, it's official," Darrell offered. "They're flying engine first and feet first to slow down."

Charlie Duke suddenly broke in over the speakers. "Columbia," Duke was calling Mike Collins in the orbiting command module. "We've lost them," he transmitted, "tell them to go to Omni."

Across the distance, everyone at Houston and at the Cape could hear Collins calling, but he also had lost his high-gain connection and was almost unintelligible over the intercom. ". . . They'd like you . . . to go to Omni . . ."

After several minutes of garbled static Charlie Duke sounded confident. "Eagle," the Cap-com transmitted. "We've got you now. You're looking good . . . Still looking good." Charlie Duke continued to advise, "Coming up on 3 minutes."

Buzz Aldrin was not looking out of his window as the gray and cinnamon lunar surface grew closer. He was too busy. Aldrin was carefully watching all of the critical display and keyboard information unfurling on his screen. Neil Armstrong was however doing some old-fashioned barnstorming math as he calculated with his wristwatch the flight angles and elapsed time passing as various craters and boulder ranges swept past. Etched onto his view port window were horizontal and vertical sighting markings to determine angel and descent rates.

Armstrong clicked his Press to Talk, "Our position checks downrange show us to be a little long of our landing site."

Charlie Duke responded solemnly through heavy static, "Roger."

Buzz Aldrin suddenly broke over the speakers and his voice sounded strong. "Abort Guidance System is showing our descent rate to be about 2 feet per second greater than primary guidance. Attitude is right down the middle."

"You know . . . I'll tell you," Armstrong's voice sounded in the distance over Aldrin's VOX, "this is much harder to do than it was . . ."

Aldrin broke in suddenly, "Keep it going."

Charlie Duke was back, "You are Go to continue powered descent. I repeat: You are Go to continue powered descent."

Chapter 101

"It was a thundering beautiful experience — voluptuous, sexual, dangerous, and expensive as hell."

— Kurt Vonnegut Jr.

29,000 Feet Above the Sea of Tranquility

"How do you look over there?" Neil Armstrong was asking Buzz Aldrin, but he was very focused on the etchings on his viewport and the passing moonscape as his partner scanned the display and keyboard information lighting up the screens.

"Okay," Aldrin paused as he scanned the redline values. "We've got a good lock on landing radar."

"We've got a lock-on?" Armstrong's hand was gripping the thruster lever as Grumman's lunar landing vehicle was rolling over to place her feet in a downward position to prepare for the lunar surface. The lunar surface, below the cinnamon crater rims and the gray boulder fields, were closer than ever before, and the scenery was moving by at a much slower rate of travel.

"Yeah," Buzz confirmed over the engine noise, "altitude lights just gone out. We're at 29,000 feet."

"Roger," Charlie Duke suddenly sounded over the firing descent engine and the constant popping of the RCS engines that kept the pitch, yaw, and roll angles in check. "Houston copies that the altitude light is out. 29,000 feet and we've got data stream again."

"Thank God for that," Les Geisman nodded as he looked first to his and then to Darrell's screen. The redline values were streaming in.

"We got the earth straight out of our front window," Aldrin's voice sounded strong over the VOX. " —about 23 degrees west of the zenith."

"We copy that." Charlie Duke sounded as if he were looking around the gathered technicians in Houston and smiling.

"We've got an alarm," Armstrong said the words calmly but there was just a touch of anxiety as he repeated: "We have a program alarm."

"It's looking good to us," Charlie Duke sounded confident.

"It's a 1202 alarm," Adrin's voice was on the VOX.

"Roger," Charlie confirmed with even more confidence, "we got you . . . We are Go on that alarm."

"Roger," Armstrong replied smoothly. "It looks good now."

"Roger," Aldrin agreed, "coming down beautifully . . . Wow! Throttle down."

"Throttle down on time," Armstrong added.

"Roger," Duke replied, "we copy throttle down."

"You can feel it in here, when it throttles down," Aldrin transmitted, "much better than in the simulator . . . Much better."

"We're at 7 minutes into the descent engine burn," Charlie Duke came over the speakers. "You're looking great to us Eagle . . . You're looking great Eagle." Charlie Duke's response was beginning to sound like an echo. "You're Go. 1202 alarm. We copy that. You're Go for landing."

Aldrin was reading the computer display numbers as Armstrong gripped the control lever over and looked out his viewport.

"35 degrees, Buzz transmitted on his VOX. 35 degrees at 750 feet —coming down at 23 feet per second. Coming down nicely . . ."

Over the speakers, everyone in the trench could hear Neil respond over his partners VOX, "Okay. Pretty rocky area," he announced matter-of-factly.

"600 feet," Buzz continued his commentary, "down at 19 feet per second."

"I'm going to take over," Armstrong made the announcement and everyone could hear the noisy popping and banging as the four sets of quad thrusters were firing to hold the
pitch, yaw, and roll in check.

The main descent engine was as steady as a jet turbine
although the pounds of fuel remaining were getting dangerously low. The two pilots were very close to the "Bingo" level when an abort would be called and the second stage of the
lunar landing vehicle would be compelled to fire and rocket
the two men back to an orbital rendezvous and a lifetime of
wondering how they had screwed up the most important
aeronautical mission in human history.

"Okay," Buzz was back on his VOX. "400 feet, down at 9
feet per second, 58 feet per second forward."

Again from his partner's voice-activated mike, Armstrong merely said, "No problem."

"300 feet," Aldrin offered. "Down one and a half per second, 47 feet per second forward. Slow it up. Ease her down
. . . 270 feet."

"Okay," Armstrong's voice was hardly audible over the
descent engine burn and the banging thrusters. "How's the
fuel?"

"8 percent," Aldrin said and the VOX crackled.

"Okay," Armstrong sounded distant. "Here's a . . . looks
like a good area here."

"I got the shadow out there," Buzz sounded very clear
over his VOX. "I can see our shadow. I can see our descent
and assent stages as a shadow—altitude and velocity lights
on . . . 220 feet and coming down nicely."

"Going to be right over that crater." Armstrong was
pitching the thrusting lever forward for a better view.

"200 feet," Buzz was a rock. "Five and one half down."

"I got a good spot," the voice from across the tight little
upright cabin reported.

"Five and one half down, 9 forward, you're looking good.
100 feet, quantity light is on. Okay, 75 feet and it's looking
good down a half . . . forward 6."

Charlie Duke suddenly intruded with a very audible
beep from Houston, "60 seconds until Bingo."

Without acknowledging the low fuel warning from Houston, Buzz Aldrin continued with his numbers, "Velocity lights on—60 feet. Two and one half down, forward 2 . . . forward 2. That's good."

"40 feet," as Aldrin's VOX continued over the speakers no one in the trench was moving or pausing to blink. "Picking up some dust," the VOX continued, "30 feet, two and one half down, faint dust shadow."

"4 forward . . . 4 forward," the voice on the distant VOX was ever so patient, "Drifting to the right a little . . . 20 feet . . . down a half."

Charlie Duke was back with another very loud beep and another stern warning from Houston, "30 seconds until Bingo!"

"Drifting forward just a little bit," Aldrin continued patiently. "That's very good. Okay contact light!"

"Shut down," Armstrong voiced the order, and his words were suddenly very clear in the following silence.

"Okay," Aldrin acknowledged. "Engine stop . . . Mode control both auto . . . descent engine command override: Off. Engine Arm: Off. 413 is in."

Charlie Duke was transmitting, "We copy you down Eagle . . ."

After a pause and as everyone in the trench and across the world was holding their breath, a loud and clear voice from the moon was reporting.

"Houston," Neil Armstrong said and he was once again on VOX, "Tranquility base here. The Eagle has landed."

"Roger Tranquility, we copy you on the ground. You've got a bunch of guys about to turn blue. We're breathing again. Thanks a lot."

At the Cape and at Houston, the control rooms erupted with delight. Everyone was suddenly standing and shouting and shaking hands, and there was a feeling of accomplishment like never before.

Meanwhile, far away from the Cape and Houston and the searching antennas at Goldstone and Honeysuckle Creek, the blue-and-white marbled sphere of the earth continued to rise and dominate the gray- and cinnamon-touched mountains on the moon. Below the rising earth, but above a debris field of otherworldly boulders, an 18-ton black-and-white landing craft with gold spider legs sat upright on a desolate and lonely vista with two little cabin lights burning brightly from a pair of triangular viewport windows. Above the blackened descent engine cone and the now blasted lunar surface, there were two words inscribed on the elevated base of the landing craft: "United States."

Behind the little lighted windows, there were two astronauts working steadily and going through their checklists as rote, but as the world now knew, it had only been twelve years since the launch of Sputnik, but now there were two pioneers getting ready for a walk, and there were going to be American boots on the moon.

Epilogue

Three Hours Before Dawn, July 30, 1969 - Titusville, Florida

When Darrell arrived at home, Audrey was waiting. She had never seen her husband so tired and excited all at once. She had watched television with the kids as Neil Armstrong and Buzz Aldrin stepped down from the ladder of the landing leg and began to walk on the moon. The kids sat for hours until well after midnight as they watched Walter Cronkite and the crew at Houston and the crew at the Cape. They never even once complained that it was taking too long, and when Armstrong touched the moon and began to walk, and Aldrin stood and took in the scenery and said "magnificent desolation," Audrey knew that this moment was one of the most special moments in history.

"Did you see everything?" Darrell asked. "Did the kids watch?"

Audrey smiled. "Of course we watched silly. Everyone, everywhere was watching."

"Did anyone call?"

"Bill Hinkle in Detroit and Will Hammond in Huntsville. They both said to tell you congratulations, and Will Hammond said to say thanks again for the sand wall at White Sands."

Darrell grinned despite his fatigue.

"How about a nightcap?" Audrey changed the subject. She knew better than to ask about the sand wall. "I know you have to get up early and go in again, but you also need to unwind."

"Yes," Darrell nodded, "a nightcap would be great."

After Audrey went to make the drinks, Darrell went into the living room where the television was still flickering and the video tape was still showing Armstrong and Aldrin standing beside the planted and unfurled American flag on the moon.

When Audrey came out of the kitchen, a bed-tousled Kath in her pajamas was with her.

"Okay daddy," Audrey said with her hands on her hips, "your daughter has an important question."

Without a pause, Darrell sat down on the sofa as Kath wiped the sleep from her eyes and curled up beside him.

"All right sweetie," Darrell looked at her closely, "what is your question?"

"Have you always been in this storm?" the little but determined voice asked, and, at first, Darrell thought she had been dreaming. "Have you always been in this storm that's always around us? Have you always been surrounded by thunder?"

About the Author

Tom Williams is a columnist and feature article writer for a series of Scripps newspapers in Southwest Florida. After having published hundreds of human interest and culinary columns, Tom's well-rounded writing career includes articles in *Paradise Magazine, Naples Daily News* and *The Marco Island Eagle*. Williams' writing has been published in Amsco school publications and in Frontier Airlines in-flight magazine, *The Wild Blue Yonder*.

After the release and publication of *Lost and Found* by Archebooks in September 2008, Tom began a promotional media tour featuring television and radio interviews across the nation. With over 25 flattering reviews to date and the endorsement of *The New York Times* Best-Selling Author Steve Alten, *Lost and Found* is proving to be *"a promising debut novel that combines a* Pirates of the Caribbean *flavor with 21ˢᵗ century technology. Perfect for sipping Piña Coladas on a tropical azure shoreline."*

With uncountable shipwrecks lost on the ocean floor, Williams' *Lost and Found* explores the possibility that a satellite orbiting above the earth could detect a super conductor such as gold under the sea, and potentially find all of the lost shipwrecks around the globe that have a wavelength signature for the incredibly precious metal. With the current price of gold, can anyone really imagine the possibilities?

Tom has appeared on Texas A&M University's KAMU television and has been interviewed as a salvage expert on "FOX and Friends" network television show. WINK television in Fort Myers, Florida, the Daily Buzz network television, and the "Joey Reynolds Show," a late night radio program on WOR radio in New York City, have interviewed Tom as well. *Lost and Found* has led Tom to Palm Springs, California, to be on KMIR television, and Tampa, Florida, for NBC Daytime T.V., the Jim Bohannon network radio show from Washington D.C., and regional radio interviews nationwide. The website for *Lost and Found* is: lostandfoundadventure.com

Flight of the Valkrye, Tom's first novel, is the prequel for *Lost and Found*, and is an epic Second World War adventure reaching into the wilds of the Carpathian Mountains of Eastern Europe, where an amazing aircraft technology is discovered that requires no fossil fuels. This technology proves to be a formula for the "George Jetson" car of the future and a global dominating success that could only be described as a checkmate move for any nation who can race to discover this lost technology.

"Surrounded by Thunder — The Story of Darrell Loan and the Rocket Men," is Williams' third work. The timeline, and all the incredible achievements of the early days of space exploration are true.

In August 2013, Tom will celebrate 29 years as a Master Merchant Marine Officer licensed by the United States Coast Guard. He is a well-traveled and veteran scuba diver specializing in shipwreck diving with a keen interest in deep-sea archeology. After decades of real life discovery above and under the water, Tom is uniquely enabled to create cause and effect scenarios where the written words paint the pictures.

Tom Williams and his writing are well represented by the Ascot Media Group in Houston, Texas, and specifically by Trish Stevens, publicist extraordinaire.

In October 2013, Williams will reach a milestone with Marriott, and specifically Marriott's Marco Island Beach Resort, where he has lived and worked on the waters surrounding Marco Island, Florida, for 35 years.

Tom Williams lives on Marco Island, Florida, with his wife Victoria and their three cats, Angel, Atticus, and Scout.

Praise for Tom Williams' previous novel, *Lost and Found*

From--Steve Alten, New York Times Best-Selling author of MEG & The SHELL GAME

"A promising debut novel that combines a *Pirates of the Caribbean* flavor with 21st century technology. Perfect for sipping Pina Coladas on a tropical azure shoreline."

From: Ron Polli Offshore Editor: Extreme Boats Magazine:

"An adrenaline filled Rollercoaster ride with a pot of gold as the prize. This is fiction, but someone in NASA could turn it into reality. All the wit and emotion of Clive Cussler's best novel. A Great read, a page-turner that you won't want to put down."

Reviewed by David C. Martin Publisher & Editor of *Health Care Weekly Review:*

"While I normally only review books related to the health care business I decided to review this novel because it promised to be intriguing. When I read it I was delighted with the pace, adventure and timeliness of this story. This is a fun adventure with all of the thrills of an Indiana Jones story."

From Larry Jewett Editor Mustang Enthusiast:

"Tom Williams really keeps the action moving. Plot twists run deep into the content and the story takes you on some interesting journeys. Just when you think you've got it all figured out, there's a surprise waiting in the next chapter. It's an intriguing read with colorful characters we've encountered in our lives. The compelling story is hard to put down and the ending is well worth the read."

From: Owen Krahn, Krahn Publishing: Idaho Senior News:

The book "Lost & Found" has more twists and turns than a corkscrew. The story is action packed and full of suspense to keep you turning one page after another. The story leads you through the deceit and treachery of the Corporate Board room to the excitement and turmoil of the Treasure Hunt. This book would make a GOOD movie. I could see the movie in my minds eye as I read the book. ENJOY!

From: Loretta Lynn Leda: Feature writer for the Orlando Sentinel, The Reporter, and the West Orange Times :

Tom Williams' first novel, Lost & Found, has an interesting storyline that threads the reader from high-tech technology and touching on current topics, to mystery and intrigue. If you enjoy pirate stories, mysteries and adventures, you will enjoy Tom's modern-day treasure hunt for survival. Well-written for his first novel. I hope to see more of his work.

From: Ralph L. Webb, Professor Emeritus Department of Mechanical Engineering Penn State University:

"This was a delightful and fun read. The book has lots of subtle humor, the characters well developed, and is a really good adventure story. Highly recommended!"

From: William Kerns entertainment reviews Lubbock Avalanche-Journal:

Kerns: Williams delivers the goods with novel Lost and Found'
"Florida-based writer Tom Williams' novel, cleverly titled "Lost and Found," will be released soon. It cannot be too soon. This extremely unpredictable adventure is guaranteed to keep even the most weary bedtime readers turning pages late at night, telling themselves "just one more chapter" over and over again."

From: Michael H. Price The Business Press Fort Worth:

"A rip-snorting thriller, rich in vicious international intrigues, corporate treacheries and old-fashioned heroic pulp-fiction gumption — all in a lifelike setting that draws as much intense momentum from its technological realism as from the life-or-death urgency of the situation. A cracking good read. Some fine reading here, and good material overall for our Lone Star Library section."

Lynne R. Christen Freelance Travel Journalist:

"Lost & Found is a fast and easy read with all of the customary components of an entertaining book...murder, mayhem, madness, mystery and romance. Just the thing to liven up a long boring airline flight or for a lazy day by the pool."

From:Brian Bandell, South Florida Business Journal:

"With treasure hunting, corporate back-stabbing and a sadomasochist on the heroes' heels, Lost & Found has everything you could ask for in a great thriller. It's a business meltdown that makes the story of Enron look tame."

From: Ed Nelson (Chicago's southernmost suburbs):

Lost & Found is a mystery adventure by an author of startling ability. I was able to put it down a few times while reading, but only because of demanding interruptions. And I resented every one of them. It's common to talk of a novel's "fast pace," but the phrase is hardly enough to describe the fevered action in Tom Williams' story. Among the *Lost & Found* cast, Billie Johnson's forthright charm acts on a pair of British expatriates the way it will on readers; she's darned near irresistible without trying to act on either of the pair. All together, you'll find *Lost & Found* just as irresistible. Williams launches a sentence that goes straight to its target -- no fooling around with "literary" pretense, just direct communication. So he presents his story simply, directly, and clearly. You'll enjoy it thoroughly"

From: Gabe VanWormer, Producer, Michigan Out-of-Doors TV:

"Tom William's has woven together a masterpiece of adventure, murder and mystery all fueled by greed. This is a must read for anybody who's ever dreamed about diving a wreck and finding a gold doubloon. William's characters will pull readers into the adventure and along on a wild ride."

From: The Weekend Star Democrat Easton Maryland and Anne Stinson: Book Critic Lost & Found:

Action packed, fast moving, engrossing thriller.
"This contemporary thriller novel has it all: high tech satellite gimmickry, brainy Brit scientists, corporate chicanery, sunken Spanish galleons, a genuine damsel in distress, arson, an earthquake and nutcase villains of multi international origins. In addition, it's action-packed, fast moving, and totally engrossing. Read it in a hammock and you won't even notice the mosquitoes."

By Lee and J.J. MacFadden Special to the Herald Courier Tri-Cities:

"Lost And Found" is an Intriguing Story and delivers thrills and surprises. Williams knows how to build suspense as well as creating believable personalities. The heroes are likable though realistically flawed, and Williams manages to make the main villain, Cedric, grow even more unlikable and unstable as the book progresses.

From: Bill Balam Senior Beacon Milford, NH :

"It has been said that the test of a good book is dreading to read the final chapter. This was the case as I was reading Lost and Found during a recent cruise to Bermuda. The excitement was riveting with no letup in sight – my type of reading. Originally, I was planning to read John Adams, a fine book but certainly lacking the adrenaline flow needed to pass the hours of a 7-day cruise.
Thank you for making the Lost and Found world part of my world. Looking forward to the next manuscript."

Rainy-day reading from the cheap seats By Lynn Elliott:

Lost & Found, by Tom Williams: Glad I "found" this one… Alright, I'm not that funny, but this book has its funny moments, especially in the interplay between a pair of expatriate Brits who fail to help a big oil company hit black gold with a satellite, but stumble on how to find real gold – hijinks and high adventure ensue.

From Bruce Von Stiers:

"Lost And Found is indeed an adventure novel worth reading. Tom Williams may truly find himself heralded among great nautical adventure novelists as Clive Cussler."

From "The Paper Book Editor" - Lee Libro:

"I found Tom William's story built up into a crescendo in all the right places and most of it was so gripping that I didn't want to put the book down. His writing has that magic quality, the ability to transport the reader effortlessly. When a writer achieves this, the reader's imagination becomes fully engaged with the story so words on a page pass hardly detectable beneath the eye. The reader no longer delineates black words on white paper, but rather internalizes full action and emotion. With *Lost and Found*, you will arrive at this true entertainment level, the apex of good fiction. On my rating scale, I give this book a 5 out of a possible 5 magic books."

The Ensign Magazine review by Don Dunlap:

Murder, Mayhem, corporate greed, and international intrigue drive this face-paced novel about scientists who develop technology to pinpoint sunken treasure ships. Hard to put down. It's the perfect book for your summer vacation.

From Stacie Hearne at Authorsreading.com :

What an extraordinary tale! One of Intrigue, sabotage, greed, conspiracy, and Gold! · Tom Williams has spun a captivating novel that is impossible to put down! You find yourself lost within the pages of this book, imagining yourself there, and secretly wishing you were! Tom Williams is truly a gifted writer. The book is a treasure to read, and I only hope you can find the gold for yourself in reading this magical novel.

From Mid-West Book Review:

"Lost and Found is an exciting novel of adventure. Highly recommended."